Periodic Table of the Elements

	IA 1	IIA 2											IIIA 13	IVA 14	VA 15	VIA 16	VIIA 17	VIIIA 18
1	¹H 1.008																	²He 4.003
2	³Li 6.94	⁴Be 9.01											⁵B 10.81	⁶C 12.01	⁷N 14.01	⁸O 16.00	⁹F 19.00	¹⁰Ne 20.18
3	¹¹Na 22.99	¹²Mg 24.31	IIIB 3	IVB 4	VB 5	VIB 6	VIIB 7	VIIIB 8	VIIIB 9	VIIIB 10	IB 11	IIB 12	¹³Al 26.98	¹⁴Si 28.09	¹⁵P 30.97	¹⁶S 32.06	¹⁷Cl 35.45	¹⁸Ar 39.95
4	¹⁹K 39.10	²⁰Ca 40.08	²¹Sc 44.96	²²Ti 47.90	²³V 50.94	²⁴Cr 52.00	²⁵Mn 54.94	²⁶Fe 55.85	²⁷Co 58.93	²⁸Ni 58.71	²⁹Cu 63.55	³⁰Zn 65.37	³¹Ga 69.72	³²Ge 72.59	³³As 74.92	³⁴Se 78.96	³⁵*Br* 79.90	³⁶Kr 83.30
5	³⁷Rb 85.47	³⁸Sr 87.62	³⁹Y 88.91	⁴⁰Zr 91.22	⁴¹Nb 92.91	⁴²Mo 95.94	⁴³Tc 98.91	⁴⁴Ru 101.97	⁴⁵Rh 102.91	⁴⁶Pd 106.4	⁴⁷Ag 107.87	⁴⁸Cd 112.4	⁴⁹In 114.82	⁵⁰Sn 118.69	⁵¹Sb 121.75	⁵²Te 126.7	⁵³I 126.9	⁵⁴Xe 131.3
6	⁵⁵Cs 132.91	⁵⁶Ba 137.34	57-70 *	⁷¹Lu 174.97	⁷²Hf 178.49	⁷³Ta 180.95	⁷⁴W 183.85	⁷⁵Re 186.2	⁷⁶Os 190.2	⁷⁷Ir 192.2	⁷⁸Pt 195.09	⁷⁹Au 186.97	⁸⁰*Hg* 200.59	⁸¹Tl 204.57	⁸²Pb 207.19	⁸³Bi 208.98	⁸⁴Po (209)	⁸⁵At (210)	⁸⁶Rn (222)
7	⁸⁷Fr (223)	⁸⁸Ra 226.03	89-102 **	¹⁰³Lr 262.11	¹⁰⁴Rf 261.11	¹⁰⁵Db 262.1	¹⁰⁶Sg 263.12	¹⁰⁷Bh 264.12	¹⁰⁸Hs 265.13	¹⁰⁹Mt 268								

* Lanthanide

⁵⁷La	⁵⁸Ce	⁵⁹Pr	⁶⁰Nd	⁶¹Pm	⁶²Sm	⁶³Eu	⁶⁴Gd	⁶⁵Tb	⁶⁶Dy	⁶⁷Ho	⁶⁸Er	⁶⁹Tm	⁷⁰Yb
138.91	140.12	140.91	144.24	144.91	150.36	151.96	157.25	158.93	162.50	164.93	167.26	168.93	173.04

** Actinide

⁸⁹Ac	⁹⁰Th	⁹¹Pa	⁹²U	⁹³Np	⁹⁴Pu	⁹⁵Am	⁹⁶Cm	⁹⁷Bk	⁹⁸Cf	⁹⁹Es	¹⁰⁰Fm	¹⁰¹Md	¹⁰²No
227.03	232.04	231.04	238.03	237.05	244.06	243.06	247.07	247.07	251.08	252.08	257.1	258.1	259.1

Gas at room temperature
Liquid at room temperature

Courtesy: James Y. Fukunaga

Number	Name	Symbol	Number	Name	Symbol
1	Hydrogen	H	55	Cesium	Cs
2	Helium	He	56	Barium	Ba
3	Lithium	Li	57	Lanthanum	La
4	Beryllium	Be	58	Cerium	Ce
5	Boron	B	59	Praseodymium	Pr
6	Carbon	C	60	Neodymium	Nd
7	Nitrogen	N	61	Promethium	Pm
8	Oxygen	O	62	Samarium	Sm
9	Fluorine	F	63	Europium	Eu
10	Neon	Ne	64	Gadolinium	Gd
11	Sodium	Na	65	Terbium	Tb
12	Magnesium	Mg	66	Dysprosium	Dy
13	Aluminum	Al	67	Holmium	Ho
14	Silicon	Si	68	Erbium	Er
15	Phosphorus	P	69	Thulium	Tm
16	Sulfur	Si	70	Ytterbium	Yb
17	Chlorine	Cl	71	Lutetium	Lu
18	Argon	Ar	72	Hafnium	Hf
19	Potassium	K	73	Tantalum	Ta
20	Calcium	Ca	74	Tungsten	W
21	Scandium	Sc	75	Rhenium	Re
22	Titanium	Ti	76	Osmium	Os
23	Vanadium	V	77	Iridium	Ir
24	Chromium	Cr	78	Platinum	Pt
25	Manganese	Mn	79	Gold	Au
26	Iron	Fe	80	Mercury	Hr
27	Cobalt	Co	81	Thallium	Tl
28	Nickel	Ni	82	Lead	Pb
29	Copper	Cu	83	Bismuth	Bi
30	Zinc	Zn	84	Polonium	Po
31	Gallium	Ga	85	Astatine	At
32	Germanium	Ge	86	Radon	Rn
33	Arsenic	As	87	Francium	Fr
34	Selenium	Se	88	Radium	Ra
35	Bromine	Br	89	Actinium	Ac
36	Krypton	Kr	90	Thorium	Th
37	Rubidium	Rb	91	Protactinium	Pa
38	Strontium	Sr	92	Uranium	U
39	Yttrium	Y	93	Neptunium	Np
40	Zirconium	Zr	94	Plutonium	Pu
41	Niobium	Nb	95	Americium	Am
42	Molybdenum	Mo	96	Curium	Cm
43	Technetium	Tc	97	Berkelium	Bk
44	Ruthenium	Ru	98	Californium	Cf
45	Rhodium	Rh	99	Einsteinium	Es
46	Palladium	Pd	100	Fermium	Fm
47	Silver	Ag	101	Mendelevium	Md
48	Cadmium	Cd	102	Nobelium	No
49	Indium	In	103	Lawrencium	Lr
50	Tin	Sn	104	Rutherfordium	Rf
51	Antimony	Sb	105	Dubnium	Db
52	Tellurium	Te	106	Seaborgium	Sg
53	Iodine	I	107	Bohrium	Bh
54	Xenon	Xe	108	Hassium	Hs
			109	Meitnerium	Mt

Courtesy: James Y. Fukunaga

INTRODUCTION TO CHEMISTRY AND THE ENVIRONMENT

Baldwin King

Wipf and Stock Publishers
150 West Broadway • Eugene OR 97401

Wipf and Stock Publishers
150 West Broadway
Eugene, Oregon 97401

Introduction to Chemistry and The Environment
By King, Baldwin
©2002 King, Baldwin
ISBN: 1-57910-905-5
Publication Date: March, 2002

To my wife, Cheryl
And our children, Bryan
Cherene and Debra

About the Author

Baldwin King received his Ph.D in physical/inorganic chemistry from the University of California, Berkeley in 1968 under the direction of the brilliant thermodynamicist, Leo Brewer. Since 1979, he has taught physical chemistry at Drew University where he is presently Professor of Chemistry. His research interests include the physical and inorganic chemistry of metal-DNA complexes.

CONTENTS

PART A - BASIC CHEMICAL PRINCIPLES

1. The Good Earth   1

     1.1   Man and His Environment   1
     1.2   Chemistry and the Environment   2
     1.3   The Earth and the Solar System   3
     1.4   The Structure and Composition of the Earth   6
         Questions   8

2. Introduction to Chemistry   10

     2.1   Matter   10
     2.2   Energy and Temperature   12
     2.3   Units and Dimensions   13
         Questions   15

3. Atoms   18

     3.1   Elements   18
     3.2   The Periodic Table   21
     3.3   Atomic Structure   22
     3.4   The Aufbau Principle   27
         Questions   30

4. Molecules   33

     4.1   Molecular Properties   33

Contents

    4.2 Bonding in Molecules 35
          Questions 40

5. Chemical Change   44

    5.1 Oxidation Numbers 44
    5.2 Chemical Equations and Stoichiometry 46
    5.3 Types of Chemical Reactions 47
    5.4 Energy Changes in Chemical Reactions 49
    5.5 Rates of Chemical Reactions 51
    5.6 Chemical Equilibrium 55
          Questions 59

PART B - AIR POLLUTION

6. The Gaseous State   62

    6.1 Properties of Gases 62
    6.2 Composition of the Atmosphere 67
          Questions 75

7. Air Pollution   77

    7.1 Polluted Air 77
    7.2 Photochemical Smog 77
    7.3 Stratospheric Ozone Depletion 82
    7.4 Acid Rain 87
    7.5 Global Warming 91
          Questions 97

PART C - WATER POLLUTION

8. Water and Aqueous Solutions   100

    8.1 Properties of Water 100

- 8.2 Aqueous Solutions 104
- 8.3 Concentrations of Solutions 105
- 8.4 Colligative Properties of Solutions 107
- 8.5 Acids, Bases and Salts 108
- 8.6 Strength of Acids and Bases 111
- 8.7 The pH Scale 113
- 8.8 Buffers 116
  - Questions 118

9. Water Pollution 121

- 9.1 Distribution of Water 121
- 9.2 Polluted Water 124
- 9.3 Phosphates and Nitrates in Water 125
- 9.4 Heavy Metals in Water 128
- 9.5 Organic Matter in Water 133
- 9.6 Organic Compounds 135
- 9.7 Organics in Water 139
  - Questions 143

10. Human Health and the Environment 146

- 10.1 Chemical Toxicology 146
- 10.2 Classification of Chemical Toxins 147
  - Questions 151

PREFACE

This little book was written to fill the void of a suitable text for a **one-semester course in chemistry and the environment for non-science majors**. It may also be useful to those persons who want to learn, on their own, something about the chemistry of the environment. The pedagogical approach is to provide the basics of chemistry but with reference to the environment as a point of interest. The text begins with an elementary, conceptual but non-mathematical treatment of chemical principles and then proceeds to apply these principles to a rational discussion of issues in air pollution (such as photochemical smog, stratospheric ozone depletion) and water pollution (heavy metals in water, waste water treatment). The last chapter deals with human health and the environment.

The text is based on a course I have taught for the past five years at Drew University, a small liberal arts university in Northern New Jersey. The course has been well subscribed by freshmen through seniors and serves as part of the science requirement of non-science majors. It is also a required course for students doing the minor in Environmental Studies. No previous knowledge of chemistry is therefore required. Given the interest of many students in the environment, the combination of basic chemistry with the environment seems to attract their attention.

The text is grounded in sound chemical principles, albeit in a conceptual framework as distinct from the numerical problem solving approach of freshman chemistry. I have found that non-science majors who have to study science do not care to deal with the quantitative aspects of science. They also seem to be more interested in applications of science, especially as they impact their everyday lives and the environment is especially topical. So I have carefully selected for discussion those environmental issues which are in the public eye such as global warming, the ozone layer, smog formation, environmental carcinogens among others.

The sequence of topics is chosen to reflect the growing trend towards science in context. The first five chapters discuss chemical principles using as far as possible environmentally important compounds such as CO_2, NO_2, O_3. This is followed by two chapters on the properties of gases and air pollution. The next two chapters discuss the properties of water/aqueous solutions and water pollution. The final chapter is an attempt to put everything in perspective by discussing human health and the environment.

Preface

I have included at the end of each chapter some suggested readings for those students who would like a more detailed discussion of the topics covered. A set of discussion-type questions ends each chapter.

I wish to thank Professor N. Bunce of the University of Guelph for his permission to reproduce two tables from his book on environmental chemistry and my colleague, Dr. J. Fukunaga, for his drawing of the periodic table. Thanks also to our former departmental administrative assistant, Jane Rhone, and to my daughter, Cherene King,, for their tremendous help in the typing of the original draft of the manuscript.

Comments on the text will be greatly appreciated.. I welcome your bringing to my attention any errors, typographical or not.

February, 2002

Baldwin King
Madison, New Jersey

PART A - BASIC CHEMICAL PRINCIPLES

Chapter 1 - The Good Earth

1.1 Man and His Environment

The world around us, our total surroundings, our environment, provides us with resources of one sort or another to sustain us human beings, as well as animals and plants. However, we have not been particularly good stewards of these resources; we have polluted the environment to the point where it has reached crisis proportions. There is an urgent need to clean up the environment and keep it clean. Cleaning up the environment is the responsibility not only of chemists, biologists and other scientists, but also of the so-called man-in-the-street as well as students. Some of the common environmental problems we face, include air pollution (photochemical smog, acid rain, global warming) water pollution, and hazardous wastes.

We want to look at these issues from a chemical point of view. In other words, we will ask and answer questions like: what is the chemical basis of photochemical smog? What substances are involved, what chemical reactions occur among these substances to finally produce the phenomenon called smog? Of course, there are also biological and medical aspects to these problems. For example, how do the components of smog affect your health? How dangerous are the levels of Pb in drinking water? So it is not really possible to separate the chemical from the biological or medical aspects in the discussion and solution of the total problem. However, for the sake of convenience, we will emphasize the chemical basis of environmental issues.

Let us begin our discussion by putting in perspective how we arrived at the present nexus between the environment and human beings. First, the environment. What is it? The environment can be defined broadly as practically everything on the earth, both living and non-living. The good earth, as you know, is one of the planets of the solar system, which, if you believe the astrophysicists, appeared some 3.5 billion years ago as a result of the Big Bang. The earth itself is a nearly spherical body, made up of concentric shells, with an inner core, an outer core (about 3000 km down), a mantle (36 km down) and a crust, (sometimes called the lithosphere). We will be mainly concerned with processes at the crust. There is also the hydrosphere (consisting of the oceans, rivers, lakes), the atmosphere

1- The Good Earth

which may be further subdivided into the lower atmosphere or troposphere (extending up to about 15 km in height) and the upper atmosphere or stratosphere (from ~15 - 50 km in height). We will be concerned with what happens in both spheres, for example, ozone formation in the troposphere (smog) and ozone depletion in the stratosphere. The biosphere is made up of all living organisms. The atmosphere, together with the magnetosphere, form an envelope around the earth, shielding it from much of the radiation and meteoric particles from outer space.

How about humans? It is believed that humans emerged as a distinct species some two million years ago in Southern Africa, first as *homo habilis*, then as *homo erectus* and finally as *homo sapiens* (*Homo sapiens* is estimated to be about 100 thousand years old). *Homo habilis* were essentially gatherers, whereas *homo sapiens* evolved into hunters as their tools became more and more sophisticated. Their discovery of fire also allowed them to convert forests into grassland; hence they engaged in horticulture (limited farming), then in agriculture.

The impact of humans on the environment was not great in early times because early humans engaged mainly in horticulture/agriculture. The main environmental consequence then was probably some depletion of the nutrients in the soil by continuous farming but the problem was not serious because farmers could simply move from one plot of land to another. Serious environmental problems, as we know them today, probably started with the Industrial Revolution which began in England about 1760. Factories proliferated due to the invention by James Watt of the steam engine. Coal was used as the fuel for the steam engine and the burning of coal had serious environmental implications, such as sulfur dioxide (SO_2) and carbon dioxide (CO_2) emissions. Waste production in general increased and not much attention was paid to the proper disposal of the waste.

1.2 Chemistry and the Environment

In this text, we hope to (1) study the basic chemical principles underlying environmental issues and (2) show how these principles can be applied to the solution of some present-day environmental problems. Take mercury pollution, for example.

Mercury (Hg) is present in small amounts in the lithosphere, hydrosphere (as methyl mercury) and atmosphere. It bioaccumulates in some organisms (e.g. shell fish in Japan) and when ingested by humans, it affects the nervous system. The maximum permissible level in human food is 0.5ppm. Mercury pollution from chloroalkali plants originated from the Hg electrodes used in the process. It was a serious environmental problem in the 1970's. However, chemical innovation led to the development of non-mercury cells in chlor-alkali plants so that the mercury problem has been largely minimized today. Another example is the use of the refrigerants called chlorofluorocarbons (CFC's). CFC's were identified as serious contributors to the stratospheric ozone depletion problem in the 1980's. CFC's were supposed to be phased out by the end of the 20th century under the terms of the so-called "Montreal Protocol". Replacement substances are currently being investigated by clever chemists at duPont and elsewhere.

Clearly, in order to apply chemical principles to the study of the environment, we have to know what those principles are. So we will begin with a study of the basic principles of chemistry and then apply them mainly to the areas of air and water pollution, although we will touch on hazardous chemicals towards the end.

Finally, chemistry (and other sciences) can play its part in protecting the environment by observation and innovation. However, it takes everyone working together to do a complete job. Public policy on the environment, for example, is often needed to protect the environment from destruction by selfish individuals or organizations. For example, the rain forests must be protected and certain automobile pollution standards must be enforced by the government. Additionally, protest groups like Green Peace have been active, for example, in trying to save the whales. We are all in this boat called earth together and we have to do whatever we can for our common survival as a species.

1.3 The Earth and the Solar System

We will first like to locate the earth within the solar system, especially in relationship to the sun, so we will begin with a discussion of the solar system in general and then move on to the earth in particular.

The earth is a part of our solar system which consists of the sun, the planets and their satellites (and other things like asteroids, comets and meteorites). The first question we need to ask about the solar system is "How did it come about?" Any theory of the origin of the solar system is highly speculative. Still, we can note the nature of the system and

1 - The Good Earth

speculate on (1) its origin, (2) how it was possibly formed some four to five billion years ago and (3) its evolution to its present status. First, let us discuss the nature of our solar system.

1. The sun (a star) contains over 99.9% of the total mass of our solar system. Its diameter is about 110 times that of the earth (about12,736 km or 8,000 miles) and its temperature is estimated at 1.5×10^7 °C.

2. The planets all revolve in the same direction around the sun in elliptical orbits and these orbits all lie in roughly the same plane.

3. The planets themselves (except Venus and Uranus) rotate about their axes in the same direction as their direction of revolution around the sun.

4. The planets form two contrasting groups: an inner group of small "terrestrial" planets (Mercury, Venus, Earth and Mars) and an outer group of large "giant" planets (Jupiter, Saturn, Uranus, Neptune) and Pluto.

5. The planets show a fairly regular spacing among themselves (Bode's Law) so that a given planet such as Saturn (1421 km from the sun) is approximately twice the distance from the sun as its next nearest planet, Jupiter (775 km).

	Sun	Mercury	Venus	Earth	Mars	Jupiter	Saturn	Uranus	Neptune	Pluto
Relative Distance	0	4	7	10	16	52	100	146	300	388

The following theories of the origin of our solar system have been suggested.

1. Kant (1755) and Laplace (1796) put forward the <u>nebula</u> hypothesis in which they visualized the original state of the sun as a rotating mass of gas (or nebula) from which, by contraction followed by increased rotation, a series of gaseous rings were disengaged by centrifugal force. The rings then condensed to

The Solar System

form the planets.

2. Buffon (1749), Maxwell, Chamberlain and Moulton preferred the collision hypothesis. They suggested the planets were formed from material torn from the sun by the impact or collision with another star. These particles solidified into so-called planetesimals which then aggregated to form an expanding planet. This theory is not widely accepted.

3. Chemical-condensation-sequence model suggests an initially hot disc of gas which subsequently cooled and solid compounds condensed out forming grains that gradually grew into planetesimals and finally planets. The high melting elements or compounds (e.g. iron) would occur in planets nearer to the sun where the temperature is quite high such as Mercury, which is in fact quite dense, whereas the original more gaseous atmosphere was most likely to occur very far from the sun in planets such as Jupiter and Saturn. This theory satisfactorily explains the present known composition of the planets which is as follows:
a. Mercury is very dense and is probably similar in composition to the earth. It has no atmosphere.

b. Venus is our nearest neighbor. Its atmosphere is very dense, consisting almost entirely of CO_2 and N_2, and its surface is obscure. Its bulk composition is probably similar to that of the earth.

c. Mars' atmosphere is rarefied but clouds and dust storms prevail. Water (H_2O) is probably present in its atmosphere because polar ice caps seem to form in winter and melt in summer. Its surface is largely composed of reddish rock and its bulk composition is like earth's except that it may be more uniform throughout.

d. Jupiter, Saturn, Neptune and Uranus have dense atmospheres consisting mainly of hydrogen (H_2), methane (CH_4), ammonia (NH_3), and helium (He), like that of the sun.

1 - The Good Earth

e. Saturn has rings of ice particles. The "giant" planets probably have interiors similar to that of earth but their surfaces seem to be covered with great thicknesses of ice. Pluto, the most distant planet, does not appear to have an atmosphere and its surface may be largely made of dark-colored rock.

1.4 The Structure and Composition of the Earth

We know today, largely from seismic data, that the earth is a highly differentiated planet consisting essentially of a liquid core (made up mainly of iron (Fe) with small amounts of nickel), a relatively thin outer crust (mainly of silicate) and a mantle (also of silicate) between the two layers. The mantle itself has a thin upper solid part which overlays a viscous asthenosphere, followed by a rigid lower mantle. Then, there is our atmosphere consisting overwhelmingly of nitrogen (78%), oxygen (21%) and smaller amounts of other gases like CO_2, H_2O and argon (Ar). The lithosphere comprises the crust and upper mantle. The other two recognized zones are the hydrosphere which comprises the discontinuous shell of water, (fresh and salt), making up the oceans, rivers and lakes; and the biosphere which is the totality of living organic matter distributed throughout the hydrosphere, atmosphere and the surface of thecrust

Summary of the Structure of the Earth

Name	Important Chemical Characteristics	Physical Characteristics
Atmosphere	N_2, O_2, H_2O, CO_2, rare gases	Gas
Biosphere	H_2O, organic matter	Solid, liquid
Hydrosphere	Salt and fresh water, snow and ice	Liquid and Solid
Crust	Normal silicate rocks	Solid
Mantle	Silicate material	Solid
Core	Fe-Ni alloy	Upper part liquid Lower part solid

The elemental composition of the crust and the earth as a whole is as follows:

Composition of the Earth

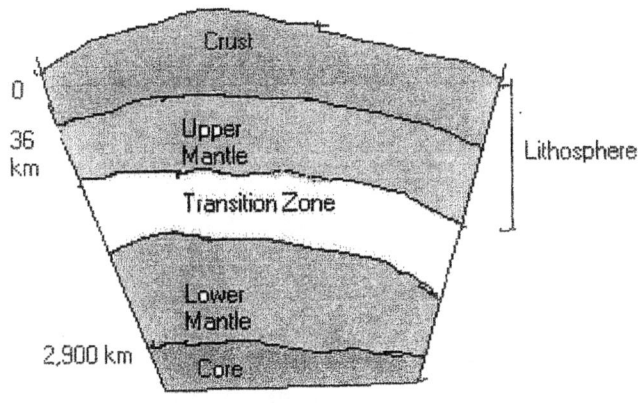

Figure 1.1   Structure of the Earth

Crust - Chemical analysis of rocks and other data show that the 8 most abundant elements in the earth's crust are in the order oxygen (O_2; 47 wt %), silicon (Si; 28%), aluminum (Al; 8%), iron (Fe; 5%), magnesium (Mg;; 4%), calcium (Ca; 3%), potassium (K; 2%), sodium (Na; 2%). These 8 elements make up nearly 99% of the total crust. Also, the earth's crust consists almost entirely of oxygen compounds mainly in the forms of silicates of Al, Ca, Mg, Na, K and Fe.

Earth - If we make certain assumptions and deductions about the bulk composition of the earth (including its interior), we find an average distribution of the elements as follows:(noting that the core and mantle make up about 99% of the earth's mass). The four most abundant elements in the earth as a whole are in the order Fe (35%), O_2 (30%), Si(14%) and Mg(13%). These four elements comprise nearly 90% of the total in the earth as a whole.

In terms of the solar system as a whole, the most abundant elements

1 - The Good Earth

are hydrogen (H_2) and helium (He) which make up the bulk of the sun.

SELECTED REFERENCES

Sections 1.1 - 1.2 Man and His Environment

1. J. M. Beard, " Chemistry, Energy and the Environment ", pp 1 - 10, Wuerz, 1995

ADDITIONAL REFERENCES

1. P. Buell, J. Girard, " Chemistry - An Environmental Perspective ", Prentice Hall, 1994
2. F. Press, R. Siever , " Earth ", Freeman, 1982

QUESTIONS

Section 1.1 - Man and His Environment

1.1 Name the main layers or zones into which the earth can be divided and show by means of a diagram their relative positions.

1.2 What is the hydrosphere? Name the other " spheres " into which the environment may be subdivided.

Section 1.3 - The Earth and the Solar System

1.3 What is a solar system? Diagram the solar system of which earth is a part.

1.4 Name the terrestial planets and say how they differ from each other in composition.

Questions

Section 1.4 - The Structure and Composition of the Earth

1.5 Name the four most abundant elements in the earth as a whole. Give their approximate percentages. Compare this composition with that of the earth's crust.

1.6 Name the two most abundant elements in our solar system as a whole. Compare this composition with that of the earth's interior.

Chapter 2 - Introduction to Chemistry

2.1 Matter

Chemistry may be defined as the science that studies matter and its behavior. In other words, it studies the composition, properties, and structure of matter as well as the energy changes that take place during the transformation of matter.

Let us begin by defining some terms. Matter is anything that has mass and occupies space. Conversely, mass is the measure of the quantity of matter in a sample of material. The air we breathe and that makes up our atmosphere is matter because it has mass because if we fill a balloon with air and put it on a sensitive balance (an instrument that measures mass) it will register a number; perhaps a small fraction of one gram but a mass nonetheless. Incidentally, note that the term "mass" was used, and not "weight" because in fact they are not the same things, even though they are often used interchangeably.

The weight of an object on a planet is the force with which it is attracted towards the center of the planet and depends on the particular planet and on the distance of the object above the surface of the planet. The weight of a body of given rest mass is six times as great on earth's surface as on the moon's surface, because the attractive force near the surface of the earth is six times as great as that near the surface of the moon. Therefore the mass, and more specifically the rest mass, of a body remains constant from planet to planet but its weight changes. We say rest mass because the mass of a body depends on its velocity and increases appreciably when its velocity approaches the speed of light. Units of mass are g (metric) and kg (SI); units of force are dynes (metric) and newtons (SI).

Classification of Matter

On the basis of its physical state, we may classify matter as a solid, liquid or gas. A solid sample has a definite volume and definite shape. It may be characterized by its density (mass per unit volume) which has a fixed value at a given temperature. (e.g. 1 mL of aluminum at 25°C and 1 atm always has a mass of 2.70 g). Also its volume or density does not vary much with changes in temperature or pressure. Most solids such as ice and aluminum are quite rigid although some are plastic, for example, clay.

At the molecular level, the atoms or molecules of a solid are very close

Classification of Matter

to each other (of the order of nm) and attract each other very strongly. These attractive forces depend on temperature. As the temperature is raised, the molecules of a solid become energized and move further apart. At a certain temperature called the melting point (0 °C for water), the solid turns into a liquid.

A liquid sample still has a definite volume but no longer has a definite shape; it takes the shape of the container. The molecules still attract each other strongly, though not as strongly as in the solid. Liquids are also not very compressible. As the temperature of a liquid is raised still further, the molecules move even further apart and more rapidly. At the boiling point (100 °C and 1 atm for H_2O), the liquid is converted into vapor (steam).

The vapor of a given sample now has no definite volume or shape. It takes the volume and shape of the container. The molecules of a gas no longer attract each other strongly. The molecules move almost independently although there is a distribution of velocities. A given substance may exist as a solid, liquid or gas depending on the temperature and pressure but whether it exists as a solid, liquid or gas at room temperature depends on the particular substance.

Let us look at another basis for classifying matter. Matter may also be divided into elements, mixtures and compounds. **Elements** are substances that cannot be broken down into any simpler substances by chemical means. Examples are sodium metal and chlorine gas. On the other hand, **compounds** are substances that are made up of two or more elements in fixed proportions by weight and can be separated into their constituent elements by chemical means. An example of a compound is sodium chloride which can be chemically broken down into sodium and chlorine.

Both elements and compounds are regarded as **pure substances** because they cannot be separated by physical means. On the contrary, **mixtures** are made up of two or more pure substances which can be separated by physical means. Mixtures are also of variable composition. For example, a mixture of sand and sugar can be separated by dissolving the sugar in water and filtering to obtain the sand. Mixtures can be further subdivided into homogeneous and heterogeneous depending on whether they are uniform throughout or not. A solution of sodium chloride in water is an example of a homogeneous mixture where a mixture of iron filings and sulfur is a heterogeneous mixture.

Here are some other definitions which apply to matter:

<u>Composition</u> refers to the identity and relative amounts of elements or

2 - Introduction to Chemistry

compounds in a sample.

Property is a characteristic of a substance which serves to identify it. Properties may be physical or chemical depending on the type of **change** that defines the property. A physical change does not result in any change in the composition of the substance. Hence a physical property of a substance is one that does not involve any change in composition of the substance; for example, density. A chemical change is one in which there is a change in the composition of the substance due to reaction. For example, methane gas (the major constituent of natural gas) burns in oxygen to form carbon dioxide and water. Hence a chemical property of methane is that it reacts with oxygen.

Properties may also be extensive or intensive depending on whether they depend on the amount of the sample or not. For example, volume and mass are extensive properties but temperature and density are intensive.

Structure shows the exact manner in which constituent particles of a substance are arranged in the substance. A molecular structure (CH_4 is tetrahedral) shows how the atoms in a molecule are oriented with respect to one another but a crystal structure (Na^+Cl^- is face-centered cubic) shows how the ions or atoms or even molecules are arranged in the crystalline solid.

2.2 Energy and Temperature

Energy is defined as the capacity to do work and work is done when a force moves a body over a certain distance. Types of energy include potential, kinetic, light, heat, nuclear, electrical, chemical. In particular, heat may be defined as energy in transit due to a temperature difference so that heat is only manifested if it flows. Units of energy are joules (J) and calories (cal); 1 cal = 4.184 J

Temperature is a measure of the 'hotness' of a substance and a temperature difference is manifested by a flow of heat.

Temperature Scales

The Fahrenheit (°F) and Celsius (°C) scales are relative temperature scales, whereas the Kelvin (K) scale is an absolute temperature scale.

Units and Dimensions

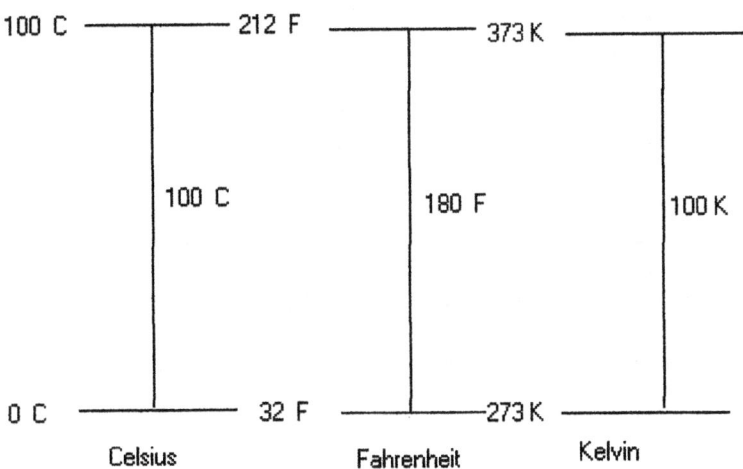

Figure 2.1 Temperature scales

The melting point of water is 32 °F or 0 °C or 273 K and the normal boiling point of water is 212 °F or 100 °C or 373 K. The scales may be interconverted by the following equations:

$$t(°F) = t(°C)(1.8°F/1.0°C) + 32°F$$
$$t(°C) = (t(°F) - 32°F)(1.0°C/1.8°F)$$
$$t(°K) = (t(°C) + 273)(1.0 K/1.0 °C)$$

2.3 Units and Dimensions

1. Systems of Units

There are two systems of units that are of particular interest to the chemist - metric and international system (SI). It was mentioned earlier that Al has a density of 2.70 g/mL at 25°C. This statement contains three units of measurement:

2 - Introduction to Chemistry

g - a unit of mass
mL - a unit of volume
°C - a unit of temperature

g (for gram) is the unit of mass in the metric system which is a decimal system (larger and smaller quantifies related by factors of 10). Prefixes are used to denote larger or smaller quantities.

$$1 \text{ kilogram (kg)} = 10^3 \text{ g}, 1 \text{ milligram (mg)} = 10^{-3} \text{ g},$$
$$1 \text{ nanogram (ng)} = 10^{-9} \text{ g}, 1 \text{ picogram (pg)} = 10^{-12} \text{ g}.$$

The SI system is derived from metric system and its unit of mass is the kg. Other quantities differ by $10^{\pm 3}$ e.g. $1 \text{ g} = 10^{+3} \text{ mg} = 10^{+6} \text{ µg}$. A unit of volume in the metric system is the cm^3 which is a derived unit obtained from the basic unit of distance (cm). Actually $1 m = 10 \text{ dm} = 10^2 \text{ cm} = 10^3 \text{ mm}$ in the metric system. The milliliter (mL) and liter (L) have been used for a long time in chemistry. The mL is essentially equal to the cm^3(cc), i.e. $1 \text{ mL} = 1 \text{ cm}^3$. Then $1 \text{ L} = 10^3 \text{ mL} = 10^3 \text{ cm}^3 = 1 \text{ dm}^3$. The basic unit of distance in the SI system is the meter (m) so that the unit of volume is m^3. We have the following relationships:

$$1 \text{ decimeter (dm)} = 0.1 \text{ m} = 10 \text{ cm} \Rightarrow 1 \text{ dm}^3 = 10^3 \text{ cm}^3 = 1 \text{ L}$$
$$1 \text{ m}^3 = 10^6 \text{ cm}^3 = 100 \text{ L}$$

The basic unit of time in both the metric and SI systems is the second (s).

2. Conversion Factors

All of these units in the metric, SI, as well as theEnglish system, are related by **conversion factors** which permit us to convert from one set of units to another. e.g. $1 \text{ km} = 0.621 \text{ mi(le)}$ and $1 \text{ h} = 3600 \text{ s}$. For example, if a car is going at 55 mph, its speed is 88 km/h or 1.5×10^{-2} mi/s, given the conversion factors above. The latter answer is written conveniently in **scientific notation** in which numbers, especially very large and very small, are written as a positive number (between one and 10) multiplied by 10^n where n could be positive or negative. The value of n is the number of places the decimal is moved either right (n<0) or left (n>0)

3. Significant Figures

The answer is also written to (what is called) 2 significant figures.

Significant Figures

The original quantity was written to 2 significant figures (55 mi/hr) representing a precise or certain number (the first 5) and an estimated or uncertain number, the second 5 (if this were an actual measurement). So, your calculated quantity must not have more significant figures than your original quantity. The rule for multiplication and division of two or more quantities is that the answer must contain a number of significant figures equal to that of the quantity with the least number of significant figures. For addition and subtraction, the answer must contain the same number of places to the right of the decimal as the quantity with the least number of places to the right of the decimal.

$$\begin{aligned} & 5.1196 \\ + \; & \underline{5.02} \\ \text{Total} \quad & 10.14 \end{aligned}$$

e.g.

SELECTED REFERENCES

Section 2.1 States of Matter

1. T. L. Brown, H. E. LeMay, B. E. Bursten, " Chemistry - The Central Science ", pp 1 - 12, Prentice Hall, 2000
2. M. S. Silberberg,, " Chemistry ", pp 1 - 13, McGraw Hill, 2000

Section 2.2 - 2.3 Energy and Temperature, Units and Dimensions

3. T. L. Brown et al., ibid., pp 12 - 28
4. M. S. Silberberg,, ibid., pp 13 - 33

ADDITIONAL REFERENCES

1. S. E. Manahan, " Fundamentals of Environmental Chemistry ", Lewis, 2001

QUESTIONS

Section 2.1 - Matter

2.1 How do we define matter? Distinguish between the mass and

2 - Introduction to Chemistry

weight of an object.

2.2　Name the three states of matter and briefly describe their differences. Give examples to illustrate your answer.

2.3　Classify each of the following as solid, liquid or gas at room temperature and atmospheric pressure. Justify your choice. (a) nitrogen (b) sodium chloride (table salt) (c) alcohol (d) aluminum.

2.4　Define the following terms. Give an example in each case. (a) pure substance (b) mixture (c) element (d) compound.

2.5　Classify each of the following as mixture, element or compound and say why. (a) carbon dioxide (b) mercury (c) beer (d) iron.

2.6　Distinguish between a physical and a chemical change and hence between a physical and chemical property. Give an example of each.

2.7　Classify each of the following as a physical or chemical change. Say why. (a) melting of snow (b) burning of natural gas (c) rusting of iron (d) dissolving of sugar in water.

Section 2.2 - Energy and Temperature

2.8　What is energy? Name four types of energy.

2.9　Distinguish between heat and temperature. Name three temperature scales and give the boiling point of water on all three scales.

Section 2.3 - Units and Dimensions

2.10　Name two systems of units frequently used in chemistry. What are the units of mass, volume, density, energy on these scales?

2.11　What are conversion factors? Write down the conversion factors to convert kilometers (km) to miles (mi) and seconds (s) to hours (h). How many mi are equivalent to 100 km? How many s are in 2 h?

2.12　How many significant figures are present in each of the following

Questions

measurements? 0.101 km and 38205 s. Rewrite these quantities in scientific notation.

Chapter 3 - Atoms

3.1 Elements

We saw earlier that elements are the simplest substances of which matter is composed. Elements are made up of **atoms** which are defined as the smallest particles which can participate in ordinary chemical reactions. Elements are represented by symbols. The symbol for an element is usually the first one or two letters of the name, for example, H for hydrogen and He for helium. The symbol represents not only the element itself but also one atom of the element. The first meaningful theory about the nature of elements and atoms was put forward by the English chemist, John Dalton in 1803. The main postulates of his theory are:

1. Elements are made up of small, indivisible particles called atoms.
2. All atoms of a particular element are alike but different from the atoms of every other element.
3. Chemical combination occurs when atoms combine in small whole number ratios to form compounds
4. During a chemical reaction, atoms are re-arranged to form new compounds; they are neither created nor destroyed.

These ideas were based on a number of experiments carried out by other chemists like Lavoisier and Proust and which were summarized in the Laws of Chemical Combination:

1. Law of Conservation of Mass (A. Lavoisier) - In a chemical reaction, matter is neither created nor destroyed
2. Law of Definite Proportions (J. Proust) - all samples of a compound contain the same elements in the same proportion by weight. For example, H_2O is always 89 % by weight of O and 11 % by weight of H.
3. Law of Multiple Proportions (Dalton) - the masses of one element combining with a fixed mass of a second element are in a small whole number ratio. For example, O combines with C to form the compounds - CO and CO_2 - in which 16 g and 32 g of O combine with 12 g of C respectively. (O ratio of 1:2)

Sub-atomic Particles

Although extremely small, atoms are themselves composed of sub-atomic particles called protons, neutrons and electrons.

Particle	Charge	Mass (amu)
Proton (p^+)	+1	1.00728
Neutron (n^o)	0	1.0086
Electron (e^-)	-1	0.000549

Protons (discovered by E. Rutherford in 1919) and neutrons (discovered by J. Chadwick in 1932) comprise the nucleus in which practically all of the mass of the atom resides. The electrons (discovered by J. J. Thomson in 1897) move in approximately elliptical orbits around the nucleus. The number of protons in the nucleus defines the atomic number Z of the atom, which is a characteristic of the atom. Because the atom is electrically neutral, Z is numerically equal to the number of electrons outside the nucleus. The sum of protons and neutrons in the nucleus is the mass number, A of the atom and is necessarily a whole number. The mass number must be distinguished from the atomic mass (in amu) which may or may not be a whole number (being very nearly, but not exactly the sum of the actual masses of all the neutrons, protons and electrons). Note that:

$$A = N_n + Z$$

where N_n is the number of neutrons in the atom. For hydrogen (protium or ordinary hydrogen) A = 1 and Z = 1, N_n = 0 and we write it as:

$$^1_1H$$

where the mass number is written as a superscript to the left of the symbol and the atomic number is written as a subscript to the left of the symbol.

Isotopes

The number of protons (the atomic number Z) in the nucleus of all atoms of a given element is the same. However, the number of neutrons and hence the mass number A may vary. Thus, there is a form of hydrogen

3 - Atoms

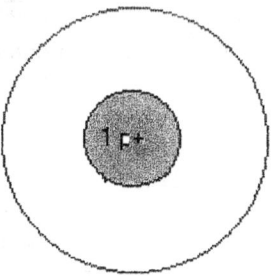

Figure 3.1 Bohr structure of the H atom

called deuterium which has one neutron and hence a mass number of 2. We write it as:

$$^2_1H = D$$

Ordinary hydrogen and deuterium are called isotopes of hydrogen because they have the same atomic number Z but different mass number A. Other examples of isotopic elements are: ^{35}Cl and ^{37}Cl; ^{117}Sn and ^{119}Sn. In fact, most of the naturally occurring elements are mixtures of several isotopes, the relative %'s of which vary from element to element. For C, the natural abundances are $^{12}_6C$ (98.89 mol %) and $^{13}_6C$ (1.11). $^{14}_6C$ is radioactive and is present in negligible amounts. ^{19}F exists as a single isotope.

Atomic Mass

The atomic mass of a given isotope of an element is its mass relative to the mass of the $^{12}_6C$ atom taken arbitrarily as exactly 12.000 amu, (atomic mass unit) so that if a given atom is twice as massive as the $^{12}_6C$ atom, its atomic mass is 24.000 amu. Note then that an amu is defined as 1/12 of the mass of a $^{12}_6C$ atom and is an extremely small unit of mass.

The Periodic Table

1 amu = 1.67×10^{-24} g $\Rightarrow$ Mass of $^{12}_{6}$C atom = 2.0004×10^{-23} g (or 2.0×10^{-26} kg)

Atomic masses and the relative abundance of isotopes may be precisely measured by an instrument called the mass spectrometer. The value of the atomic mass of each element listed in the periodic table is the weighted average of the masses of its naturally occurring isotopes. For carbon, this mean atomic mass works out to be 12.011 amu.

3.2 The Periodic Table

Dimitri Mendeleev (1869) arranged the elements (63 known then) in increasing order of atomic masses and observed that elements with similar physical/chemical properties appeared at regular intervals (this is one version of the periodic law).

He then arranged the elements so that those with similar chemical properties occurred vertically in columns called groups; the resulting rows of elements were called periods. He left gaps in the table where elements were missing at that time. There are now 110 elements, 92 of which occur naturally. Each element has its own symbol, its atomic number and average atomic mass. The elements are broadly classified as **'representative'** or 'main group' elements and **'transition'** elements. After we have talked about the electronic configuration of atoms, we will return to the periodic table and try to relate the behavior of the elements to their position in the periodic table.

The elements in the periodic table may also be classified as metals and non-metals as follows:

Metals:
    (a) Occur on the left (2/3) of the periodic table
    (b) Have a 'metallic' luster (shiny) and are all solids except Hg
    (c) Have good thermal and electrical conductivity
    (d) Are generally ductile (drawn into wire) and malleable (hammered into sheets)
    (e) Generally have high melting points (tungsten melts at 3400°C and is used in light bulbs)
    (f) Have chemical properties which vary from very reactive (Na) to non-reactive (Au)

3 - Atoms

Non-metals:

    (a) Occur on the right (1/3) of the periodic table
    (b) Exist as brittle solids (S, I, C graphite) or gases (N_2, O_2, Cl_2); bromine is a liquid at room temperature
    (c) Have very little conductivity (electrical and thermal) excepting graphite
    (d) Have chemical properties which vary from very reactive (Cl_2) to non-reactive (Ne)

Metalloids:

These are elements that have properties intermediate between metals and non-metals; examples are B, Si, Ge, As. They tend to be semi-conductors. They represent the gradual transition from metal to non-metal either across a period, for example, Al (metal), Si (metalloid), P (non-metal) or down a group, for example, C (non-metal), Si, Ge (metalloid), Sn, Pb (metals).

Transition Elements:

These are located in the center of the periodic table between the two blocks of representative elements. They are all metals (thus lying to the left of the metalloids). They are not very reactive.

Post-transition Elements

These are metals that occur immediately after a row of transition metals (e.g. Sn, Pb) and are quite different in properties to those metals preceding the transition metals.

3.3 Atomic Structure

Prior to 1913, a few scientists, notably Ernest Rutherford, had speculated about the structure of the atom or how the electrons and protons were arranged with respect to one another. Based on data obtained from studies of the hydrogen lamp, the Danish physicist Niels Bohr in 1913 put forward his famous theory of the hydrogen atom which is the simplest atom and consists of a single proton and a single electron.

In non-technical language, Bohr's theory of the H atom (1913) can be

Atomic Structure

summarized as follows:

1. The electron revolves around the proton(nucleus) in circular orbits .
2. Only certain orbits, defined by their radius, were allowed. These orbits were labelled by n where n is an integer (1, 2, 3, etc) called a quantum number.
3. The energy of the electron in an allowed orbit had a fixed value (that is, an electron does not absorb or emit energy while in a given orbit). Hence only certain energies were allowed and again given by the value of the quantum number, n.
4. The electron moved from one orbit to another by absorbing or emitting energy.

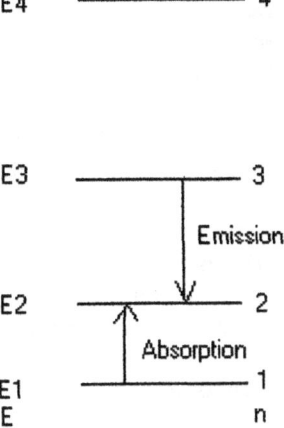

Figure 3.2 Some energy levels of the H atom

Bohr's theory was successful in explaining certain properties of the H atom such as the radiation emitted by an H lamp and so lent some validity to the suggested structure of the H atom.

Bohr's theory could not explain some of the intricate details of the H

3 - Atoms

atom and more complex atoms. Sommerfeld (1916) refined Bohr's theory by assuming elliptical orbits for the electron in the H atom. Schrödinger (Austrian, 1928) introduced other quantum numbers besides n, to describe all the possible properties of the electron in the H atom by writing and solving complex mathematical equations (the discipline of wave or quantum mechanics). The four numbers are n (the principal quantum number), l (the angular momentum quantum number), m_l (the magnetic quantum number) and m_s (the spin quantum number). The assumption is that if we know these four quantum numbers for each electron, then we know everything about the electron, its energy, momentum, magnetism. (These numbers are akin to your home address - street, city, state, country). For the moment we are mainly concerned with n and l which furthermore we will regard only as labels for the electron and its energies.

As in the Bohr theory, n is a positive integer having possible values 1,2,3,∞. l is a positive integer which can have values, 0,1,2,...n-1 and when combined with n to describe the energy of the electron cannot be greater than n-1. What we mean is, if n=1, then the only possible value of l the electron can have is 0 (i.e. n-1) and we can describe the behavior of an electron as 1,0. For simplicity we designate l=0 by the letter s and so the conventional description of the electron is 1s.

If the energy state of the electron is such that n=2, then the possible values of l are 0,1 (n-1) and we have states described as (2,0) or (2,1). The first is again conventionally denoted as 2s. For l =1 we use the letter p and the second energy state is denoted as 2p. If n=3, l can be 0,1,2 and we can have states described as 3s,3p,3d (for l = 2). For n=4, l is 0,1,2,3 and the energy states are 4s,4p,4d,4f (l =3) What is the ordering of these energy states (levels), that is, as the electron in the H atom absorbs more and more energy, what would be its successive energy states. The order is:

1s<2s<2p<3s<3p<4s<3d<4p<5s<4d<5p<6s4f<5d<6p<7s<5f

Finally, n defines a shell. The n=1 shell is sometimes called the K shell, n=2 called the L shell and so sequentially for M,N,O,P (Note that the word 'shell' has little to do with spatial arrangement; it only tells you the relative energy of the electron). Again l defines a sub-shell so that K shell (n=1) has only one sub-shell (s, sub-shell), the L shell (n=2) has two sub-shells (s,p). The M shell (n=3) has 3 sub-shells (s,p,d). In short the number of sub-shells in a shell is equal to the value of n for that shell.

Because there is only one electron in the H atom, and in its ground (lowest energy) state, it has the energy, 1s, we write: $1s^1$

Electronic Configurations of Atoms

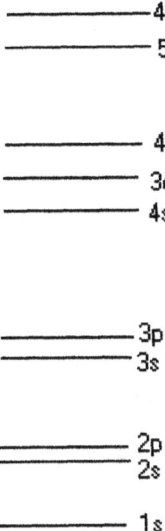

Figure 3.3 The ordering of energy levels in the H atom

to represent the ground state electronic configuration of the H atom.
Let us turn now to the next simplest atom $^4_2$He, which contains a positively charged nucleus containing two protons, and two neutrons and two electrons on the outside. What are the possible values for the energies of each electron? Fortunately the possible energy states are essentially the same as for H atom, but now we have to worry about two electrons and their combined energies. One question we can resolve now is the ground state configuration of He. It turns out that both electrons can have the energy described by 1s and so we can write for the He atom:

$$1s^2$$

We now come to Li, $Z = 3$. What are the possible ground state configurations?

$$1s^3 \text{ or } 1s^2\,2s^1$$

3 - Atoms

The first configuration is unacceptable because what we find from quantum theory is that, for a many electron atom (a) the maximum number of electrons for a given value of n is given by $2n^2$ as the following table shows:

n	1	2	3	4
#e	2	8	18	32

so that only 2 electrons in a given atom can have energies of the K shell (in which n =1)
(b) the s sub-shell can be occupied by a maximum of two electrons. In general, for a given value of l, the maximum number of electrons is $2(2l+1)$. Thus the p sub-shell has 6, the d has 10 and the f has 14. These facts are coupled with the observation that the sequence of filling up the shell and sub-shells with electrons to obtain the ground state configuration is in the order of increasing energies of the orbitals. (orbital is another word for energy state of the electron.) Therefore, the ground state configuration of Li is $1s^2 2s^1$. Note that the sum of the superscripts gives the atomic number or total number of electrons in the atom.

In like manner, we can write the ground state electronic configurations of the next 7 elements as in the follow table:

Element	Atomic Number	Electron Configuration
H	1	$1s^1$
He	2	$1s^2$
Li	3	$1s^2 2s^1$
Be	4	$1s^2 2s^2$
B	5	$1s^2 2s^2 2p^1$
C	6	$1s^2 2s^2 2p^2$
N	7	$1s^2 2s^2 2p^3$
O	8	$1s^2 2s^2 2p^4$
F	9	$1s^2 2s^2 2p^5$
Ne	10	$1s^2 2s^2 2p^6$

These ten elements represent the first and second short periods of the periodic table.

Valence and Core Electrons

The electrons in an atom may be classified further into two types-core and valence electrons. Valence electrons have energies usually represented by the largest value of n, that is, the electrons in the outermost shell (and includes all the associated sub-shells). All the other electrons are called the core electrons. We will see later that it is the valence electrons which are involved when the atoms bond together and which, therefore, determine the chemical properties of the element.

EXAMPLE:

$$Be\ 1s^2 \| 2s^2 \to 2\ \text{valence}\ e^-,\ 2\ \text{core electrons}$$
$$O\ 1s^2 \| 2s^2\ 2p^4 \to 6\ \text{valence}\ e^-,\ 2\ \text{core electrons}$$

Noting that He: $1s^2$ has 2 electrons, we sometimes regard the next 8 atoms as being built on a He core and write the configurations as:

Li: [He]$2s^1$, Be: [He] $2s^2$, F:[He] $2s^2 2p^5$ to emphasize the valence electrons.

3.4 The Aufbau Principle

In building up the periodic table so far, we have been using what is called the aufbau or building-up principle which assumes that the incoming electron occupies the lowest energy level available to it. Let us now proceed to write the electronic configurations of the rest of the elements of the periodic table beyond Ne ($12^2 2s^2 2p^6$) First we note the order of energies again:

$$1s<2s<2p<3s<3p<4s<3d<4p<5s<4d<5p<6s<4f<5d<6p<7s<5f$$

This order can be remembered using the mnemomic shown in Figure 3.4.

At high atomic number, the order is changed somewhat between 4s and 3d. For example, at Cu (Z = 29), 3d <4s unlike the situation with Ca (Z = 20) because half-filled and filled orbitals have extra stability (lower energy)

We have considered periods 1 and 2 already in going from H to Ne. The third period starts with Na ([Ne]$3s^1$) and ends with Ar([Ne]$3s^2 3p^6$. The fourth period begins with K : [Ar] $4s^1$, then comes Ca: [Ar] $4s^2$. The 3d level begins to fill up at Sc: [Ar] $4s^2 3d^1$ and ends 9 elements later with Zn: [Ar] $4s^2 3d^{10}$. NOTE: Cu is better written as [Ar]$4s^1 3d^{10}$. These 10 elements are called transition or d-block elements and they are all metals.

The next sub-shell to fill is the 4p so that Ga is [Ar] $4s^23d^{10}4p^1$. The fourth period ends with Kr: [Ar] $4s^23d^{10}4p^6$ for a grand total of 18 elements in period 4 (called a long period compared to short periods 2 and 3 which contain 8 elements). Note that, in period 4 there are 8 non-transition elements, sometimes called main group or representative elements corresponding to the 8 groups we met previously in periods 2 and 3.

The 5th period begins with the alkali metal Rb:[Kr]$5s^1$ then comes Sr: [Kr] $5s^2$. Then, there are 10 transition or d-block metals beginning with Y: [Kr]$5s^24d^1$ and ending with Cd:[Kr]$5s^24d^{10}$. Six more elements and the 5th period ends with Xe: [Kr]$5s^24d^{10}5p^6$. Again, there are 18 elements in period 5; namely 8 main group and 10 transition elements.

Period 6 starts with Cs and Ba: [Xe] $6s^2$. The next level up is 4f so La is [Xe] $6s^24f^1$. The next 13 elements fill up the 4f level ending withYb [Xe]$6s^24f^{14}$. These 14 elements (La to Yb) are called collectively the lanthanides and is the first set of inner transition or rare earth or f-block elements.

Lutetium, Lu starts the third transition metal series with the configuration Lu: [Xe]$6s^24f^{14}5d^1$ and Hg ends it [Xe]$6s^24f^{14}5d^{10}$. Period 6 has 32 elements all together: 8 main group, 10 transition and 14 inner transition. The 7th period starts with Fr and Ra and also has 14 inner transition elements called the actinides beginning with Ac: [Rn]$7s^25f^1$ and ending with No : [Rn]$7s^25f^{14}$. Lawrencium, Lr: [Rn] $7s^24f^{14}6d^1$ starts the next transition series. We are up to element 110.

<u>Chemical Periodicity Revisited</u>

Let us rewrite the electronic configurations of both Li and Na:

Li : [$1s^2$]$2s^1$ and Na: [$1s^22s^22p^6$] $3s^1$

to emphasize the core (He = $1s^2$, Ne = $1s^2 2s^2 2p^6$) and valence electrons (ns^1). Both Li and Na have one s electron in their valence shell and this accounts for the fact that many of their chemical properties are similar. e.g. they form monoxides (M_2O) and displace H_2 from H_2O. They are both called reducing agents because they give up their valence electron easily. These two elements form part of a larger group of metals labeled Gp 1A (a main group in which s and p orbitals are being filled). In Gp VIIA

Chemical Periodicity

similarly, F :[He]2s^{2}2p^5 and Cl :[Ne]3s^{2}3p^5 have similar properties. For example they are both oxidizing agents because they readily accept electrons from metals like sodium, Na.

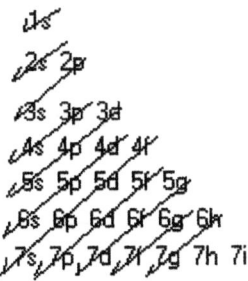

Figure 3.4 A mnemonic used in the ordering of energy levels in an atom

$$X_2 + 2\,e^- \rightarrow 2X^-$$

It is no coincidence that if we arrange the first 18 elements in increasing order of their atomic number, we find that the chemical properties of the elements vary systematically with their positions in the periodic table, or as we sometimes say, are a periodic function of their atomic number. This is a statement of the Periodic Law: that elements with similar chemical properties fall at definite intervals in the periodic table. The phenomenon is called chemical periodicity.

Lewis Dot Structure of Atoms

A convenient way to represent valence electrons in an atom is by Lewis

3 - Atoms

electron dot structures, in which, an imaginary square is drawn around the symbol for the element and each side is filled singly first, then doubly (in anticlockwise direction preferably) according to the number of valence electrons, with He being the exception. Sometimes they are drawn to reflect filled or unfilled s,p orbitals.

H· He: Li· Be· ·B· ·C· ·N· :O· :F· :Ne:

This representation anticipates the formation of two electron bonds between two atoms in a molecule.

SELECTED REFERENCES

Section 3.1 - 3.2 Elements, Periodic Table

1. T. L. Brown, H. E. LeMay, B. E. Bursten, " Chemistry - The Central Science ", pp 35 - 46, Prentice Hall, 2000
2. M. S. Silberberg,, " Chemistry ", pp 40 - 60, McGraw Hill, 2000

Section 3.3 Atomic Structure
3. T. L. Brown, ibid., pp 187 - 217
4. M. S. Silberberg,, ibid., pp 267-281, 299-307

ADDITIONAL REFERENCES

1. S. E. Manahan, " Fundamentals of Environmental Chemistry ", Lewis, 2001

QUESTIONS

Section 3.1 - Elements

3.1 Define an element. According to Dalton, what are the basic building blocks of elements?

3.2 State the Law of Definite Proportions and which postulate in Dalton's atomic theory explains it.

3.3 Name the three sub-atomic particles of interest to chemists and say how they differ.

3.4 Define atomic number and mass number. What are isotopes?

3.5 How many protons, neutrons and electrons are present in the following isotopes? (a) Sn-119 (b) Ti-48 (c) C-14.

3.6 Distinguish between the atomic mass and mass number of an isotope. Give an example.

Section 3.2 - The Periodic Table

3.7 What is a group and a period with reference to the periodic table. List one group and one period.

3.8 Give three characteristics of metals and non-metals and give two examples of metals and non-metals.

3.9 State the group and period to which the following elements belong and identify each element as metal, non-metal or metalloid. (a) sulfur (b) zinc (c) neon (d) chlorine.

3.10 What are transition elements and post-transition elements? Name the transition and post-transition elements in the following list. (a) mercury (b) lead (c) cadmium (d) arsenic.

Section 3.3 - Atomic Structure

3.11 Name the four quantum numbers that define the behavior of an electron in an atom. What are the possible values of the four quantum numbers for a 3p electron?

3 - Atoms

3.12 What is meant by the electronic configuration of an atom? Write down the complete electronic configuration of the following atoms. (a) carbon (b) oxygen (c) phosphorus (d) iron.

3.13 What are valence electrons? By writing down their electronic configurations, determine the number of valence electrons present in the following atoms. (a) sodium (b) chlorine (c) calcium (d) silicon.

3.14 Write down the Lewis dot structure of the following atoms. (a) carbon (b) nitrogen (c) magnesium (d) fluorine

Chapter 4 - Molecules

4.1 Molecular Properties

A molecule of a substance is defined as a group of atoms (same or different) which is held together by chemical bonds and which functions as an individual entity. We sometimes say that the molecule is the smallest, individual species capable of independent existence. The oxygen molecule is O_2 (diatomic), the water molecule is H_2O (triatomic), the NH_3 is tetraatomic. Beyond diatomic, we use the word polyatomic. Note that the molecules of the noble (rare) gases like He are monatomic.

Molecular Formulas

A molecule is described by a molecular formula which shows the actual number of atoms of each element in a molecule of the substance, e.g. benzene is C_6H_6, water is H_2O. The exact geometrical arrangement of the constituent atoms is given by the structural formula. Two different substances may have the same molecular formula but different structural formulas. They are called isomers, for example, dimethyl ether and ethanol of molecular formula, C_2H_6O. A third type of formula is the empirical formula which shows only the ratio of the number of atoms of each element in the compound, for example, CH for both C_6H_6 and C_2H_2. The empirical formula is related to the molecular formula by a small whole number (2 in the case of acetylene and 6 in the case of benzene).

Molecular Mass

A molecule is also described by its molecular mass (MM) which is the weighted sum of the atomic masses of its constituent elements. It is the mass of the molecule relative to ^{12}C as 12.000 amu.
For example, let us determine the molecular mass of benzene, C_6H_6. First we have the following conversions or equivalencies:

$$1 \text{ atom C} = 12.011 \text{ amu} \quad 1 \text{ atom H} = 1.008 \text{ amu}$$

Then: 6 atoms C x $\dfrac{12.011 \text{ amu C}}{1 \text{ atom C}}$ = 72.066 amu C

4 - Molecules

$$6 \text{ atoms H} \times \frac{1.008 \text{ amu H}}{1 \text{ atom H}} = 6.048 \text{ amu H}$$

Total = MM = 78.114 amu

Note that the molecular mass is quoted in amu. However, this is a very small quantity and we often quote the molecular mass in grams which for benzene is 78.114 g.

The Mole

The mole is defined as the quantity of substance which is numerically equal to the atomic mass of an atom/element or the molecular mass of a molecule/compound. It may be expressed in any convenient unit of mass such as g, kg, lb, etc., then we talk of the g. mole, kg.mole, lb.mole, etc. We shall confine our discussion to g.mole so that we shall omit the g in future discussion and further abbreviate it as mol. Then 1 mol O_2 molecules = 32.0 g O_2 molecules. However, 1 mol O atoms = 16.0g O atoms, so that 1 mol O_2 molecules = 2 mol O atoms (32.0g).
1 mol H_2O contains 2 mol H atoms (or 1 mol H_2 molecules or 2.016 g hydrogen) and 1 mol O atoms (or ½ mol O_2 molecules or 16.0 g oxygen). So, the molecular formula of a substance not only represents 1 molecule but also 1 mol of the substance.

An important finding by Avogadro was that 1 mol of any substance contained the same number ($N = 6.02 \times 10^{23}$) of atoms or molecules (or formula units for ionic compounds). So we have the conversion:

$$1 \text{ mol} = 6.02 \times 10^{23} \text{ molecules (atoms, particles)}.$$

Let us try a simple calculation involving moles.

How many mols and molecules of N_2 are there in 56.0 g nitrogen gas?

$$56.0 \text{ g } N_2 \times \frac{1 \text{ mol } N_2}{28.0 \text{g } N_2} = 2.00 \text{ mol } N_2$$

Then $2.00 \text{ mol } N_2 \times \frac{6.02 \times 10^{23} \text{ molecules } N_2}{1 \text{ mol } N_2} = 1.20 \times 10^{24}$ N_2 molecules

Ionic Bonding

4.2 Bonding in Molecules

Ionic Bonding

Consider the reaction between sodium metal and chlorine gas to form sodium chloride.

$$Na_{(s)} + \tfrac{1}{2}Cl_{2(g)} \rightarrow NaCl_{(s)}$$

The reaction may be visualized as taking place in several steps:

$Na \rightarrow Na^+ + 1\ e^-$ (loss of electron by Na)
$Cl + 1\ e^- \rightarrow Cl^-$ (gain of electron by Cl)
$Na^+ + Cl^- \rightarrow Na^+Cl^-$ (formation of sodium chloride)

In NaCl, the two ions are held together by electrostatic (Coulombic) attractions between opposite charges. This is the ionic bond in which we assume a complete transfer of electron(s) from a metal (Na) to a non-metal (Cl).

Characteristics of ionic compounds are:
(a) they are formed between elements at the extreme left and extreme right of the periodic table
(b) they have high melting points (the melting point of NaCl is 801 ° C)
(c) their solutions in water are electrically conducting; hence they are called electrolytes.

Ionic compounds are named by naming the metal first, then the non-metal which is given the ending 'ide'. For example, MgO is magnesium oxide.

Covalent Bonding

Let us look at the structure of the H_2 molecule which is made up of two H atoms. The electronic configurations and Lewis dot structure are:

$$H\bullet \qquad \bullet H$$
$$1s^1 \qquad 1s^1$$

in which each atom has one 1s electron. The question is: how are the two electrons redistributed when the two atoms come together to form the molecule and what is the driving force for bond formation. We saw that

4 - Molecules

the He ($1s^2$) and Ne ($1s^2 2s^2 2p^6$) configurations represented extra stability and so because systems move towards greater stability (lower energy), **the driving force for molecule formation is the attainment by all the atoms of the stable electronic configurations of He, Ne and the other noble gases** (by any means necessary). One possibility is for the two H atoms to share the pair of electrons so that each H atom has effectively two electrons (He configuration).

$$H \cdot\cdot H$$

This type of two electron bond, usually represented by a line through the two electrons, was first visualized by G.N. Lewis (an American chemist) and is called a covalent bond and the structure is called the Lewis structure. This approach is also called the valence bond approach.

Another possibility is for one H atom to transfer its electron to the other H, forming in turn H^+ and H^-. The two ions would then be held by an ionic bond due to electrostatic attraction between two opposite charges. The second possibility has only a very small but finite chance of occurring, partly because of the difference in stability of H^+ (unstable configuration) and H^- (stable configuration) Both ions do exist in other systems (H^+ in high-energy systems and H^- in NaH) but together they do represent a low probability of existence. Thus, we normally describe the H_2 molecule as being covalently bonded and we say that the bond has essentially 100% covalent character. However, we say more accurately that the bond has a very small but finite (5%) ionic character, so that 100% covalent (and 100% ionic character) represent extremes, with the bonding between any two atoms being made up of more or less of each kind.

The H_2 covalent bond is also called a single bond because the two atoms are held together by a single pair of electrons. The bond is strong as measured by the bond energy (436 kJ mol^{-1}) which is defined as the energy necessary to break the bond between the two atoms in the molecule.

Let us now turn to the structure of the Cl_2 molecule. Again, writing the Lewis dot structure for each Cl atom, we see they can both attain the Ar configuration by sharing one pair of electrons

$$:\overset{..}{Cl}\cdot\cdot\overset{..}{Cl}:$$
$$\underset{..}{}\underset{..}{}$$

Notice however that each atom still has three pairs of electrons which

Covalent Bonding

did not take part in the bonding. These are called lone pairs or non-bonding electrons. These lone pairs may become involved in the reactions of some molecules.

H_2 and Cl_2 are called homonuclear diatomic molecules because they are made up of the same two atoms. One consequence of this is that the two electrons forming the single covalent bond are shared equally between the two atoms, meaning that, on the average, both electrons spend as much time near one atom as the other. The molecule is then said to be non-polar. We can now define the bond length as the distance between the nuclei of covalently bonded atoms. Further for homonuclear diatomics, one-half the bond length is taken as the covalent radius of each atom.

If we now look at the HCl molecule, again we can form a covalent bond as follows:

$$H\cdot \overset{..}{\underset{..}{\cdot Cl:}}$$

However, the Cl atom has a strong affinity for electrons. As a result, the two electrons in the bond are not equally shared. The electron pair is, on the average, closer to the Cl atom and we express this by saying that there is a separation of charge with a slight excess of negative charge (δ-) on the Cl atom and a correspondingly slight equal positive charge (δ+) on the H atom, represented as

$$\overset{\delta^+}{H} - \overset{\delta-}{Cl}$$

The bond is then said to be polar and the process is called bond polarization.

We can now introduce the concept of electronegativity (EN). The atom which exerts the stronger attraction for the electron pair of a covalent bond is said to be more electronegative. The polarity of the bond is related to the difference in EN's of the two atoms. It is not possible to determine the absolute value of the electronegativity of an atom. However, by assigning an arbitrary value to one or more atoms, we can develop a relative scale of electronegativities. In the Pauling scale, the F atom is set at 4.0 and Li at 1.0 and the values for all other atoms are determined from measurement of certain physical properties (e.g. dipole moment) of a molecule containing the atom of interest.

On the Pauling EN scale, C = 2.5 and H = 2.1 and we will predict that C and H form a covalent bond in which the electron pair is shared more or less equally. The most electronegative atoms of common interest are F

4 - Molecules

(4.0), O (3.5), N (3.0), Cl (3.0).The greater the difference in electronegativity between the two atoms, the more polar the bond. With HCl, the electronegativity difference is relatively large (0.9) and indicates that the HCl bond is very polar, as we saw earlier.

Let us continue to look at the formation of compounds between H and other elements. With oxygen we have the formation of H_2O which we may visualize as :

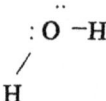

in which O forms two single covalent bonds with 2 H atoms, thus forming the octet while each H atom forms its duet.We also have two non-bonding or lone pairs on O. Note the disposition of the H atoms relative to O. This is deliberate indicating that the shape of the H_2O molecule, that is, the relative positions of the atoms, is not linear but bent. The angle formed by the two covalent bonds is called the bond angle which, for H_2O, is experimentally found to be 104°. Each OH bond in H_2O is polar and the molecule as a whole is also polar. Thus, the H_2O molecule acts as a single molecular dipole.

The polar nature of a molecule has important consequences. Solids have higher melting points and liquids have higher boiling points than non-polar molecules with the same number of electrons. H_2O's boiling point of 100°C is extremely high [relative to H_2S (-30°C)] Polar molecules also dissolve polar molecules and non-polar dissolves non-polar, so we will predict that HCl gas dissolves readily in H_2O to form a solution called hydrochloric acid.

When we look at the NH_3 molecule we find three single covalent bonds, hence three bonding pairs of electrons and one lone pair.

$$\overset{..}{H-N-H}$$
$$|$$
$$H$$

The shape of the NH_3 molecule is pyramidal with the N atom at the apex and the HNH bond angle is 107° (very close to H_2O). Again the molecule is polar due to the unsymmetrical distribution of charge and we will predict that NH_3 is soluble in water.

Covalent Structures

With methane, four covalent bonds and no lone pairs make up the Lewis structure. The shape of the CH₄ molecule is tetrahedral, the H atoms being disposed towards the corners of a tetrahedron. The HCH bond angle is 109°.

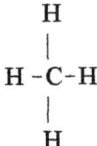

Unlike NH₃ and H₂O, CH₄ is non-polar since the CH bond is essentially non-polar and also the bond polarities cancel each other out. The forces holding non-polar molecules like CH₄ together, in the condensed state, are called van der Waals or London forces which arise from the instantaneous polarizations of the molecules due to oscillations of the electrons about their mean positions; thus, a weak attractive force is created. The noble gases are held together almost exclusively by van der Waals forces.

To summarize this section, the number of single covalent bonds formed between H and one atom of C, N, O and Cl is equal to 8 minus the number of valence electrons or the number of unpaired electrons in the electron dot structure of the atom, i.e. 4,3,2 and 1 respectively. Then by sharing with the H atom, each C, N, O, Cl atom attains a noble gas configuration (octet of outer shell electrons or the octet rule).

Some recognizable properties of covalent compounds are:

(a) they are often formed between non-metals (although metals do form covalent bonds)
(b) they are usually gases, liquids, or low melting solids at room temperature
(c) their solutions in H₂O are usually electrically non-conducting (non-electrolytes).

Let us look more closely at the bonding in a few more diatomics, for example O₂, and N₂. We can try writing the electron dot structures and, as before, try to form a single covalent bond, but notice the octet rule is not obeyed for each O atom unless the two O atoms share another pair of electrons. Thus, we have two bonding pairs between the atoms (constituting a double bond) and two lone pairs. The double bond is much stronger than the single bond (BE = 497 kJ/mol for O₂) and the bond

4 - Molecules

length is correspondingly shorter.

$$:\!\ddot{O} = \ddot{O}\!:$$

In the N_2 molecule, two N atoms are held together by a triple bond. The triple bond is extremely strong (BE = 946 kJ/mol) and partly explains the relative unreactivity and stability of N_2 gas.

$$:\!N \equiv N\!:$$

Here are some general rules for writing Lewis structures of covalently bonded molecules in which all the atoms obey the octet rule.

(1) Determine the total number of valence electrons (A) by counting the contribution from all the constituent atoms (in N_2, this number is 10).

(2) Use the octet (or duet) rule to determine the total number of electrons required for saturation (S) of all the atoms. (in N_2 this is 16).

(3) Take the difference between (1) and (2) (S - A) which gives the number of electrons to be shared (6 in the case of N_2)

(4) Choose the most probable central atom in a polyatomic species (choice may or may not be unique) and then arrange the electrons to be shared in pairs between atoms, doubling up and tripling up if necessary (hence N≡ N) so that each atom obeys the octet (duet) rule

(5) Complete the structure by drawing in all valence electrons.

Let us apply these rules to CO_2.

$$CO_2 = C + 2O$$

1. Total number of valence electrons(A) =1(4) + 2(6) = 16
2. We need S = 3x8 = 24 electrons to satisfy the octet rule.
3. The difference S-A is 8 electrons.
4. Taking the C atom as the central atom, we can deduce there are 2 C-O bonds in the molecule. We need four pairs of electrons and four bonds ⇒ two double bonds
5. The complete structure is:

Lewis Structures

$$:\!O = C = O\!:$$

Note that the CO_2 molecule is linear as shown and hence non-polar, even though the individual C-O bonds are polar.

We have dealt so far with neutral molecules but we can extend the rules for writing covalent structures to polyatomic ions with only slight modifications (polyatomic ions are species that have a net charge but contains covalent bonds). Let us take the ammonium ion (NH_4^+) which is formed when NH_3 dissolves in water. The reaction is:

$$NH_{3(g)} + H_2O_{(l)} \rightleftharpoons NH_4^+{}_{(aq)} + OH^-{}_{(aq)}$$

The modification to our rules occurs in rule (1) where, to get the total number of valence electrons for cations, we first consider the number of valence electrons assuming neutrality, then we subtract off a number of electrons equal to the charge on the cation. (So, here we subtract $1e^-$) If it is an anion, we add a number of electrons equal to the charge on the anion (for CO_3^{2-}, we add $2e^-$); rules (2), (3), (4), and (5) are then applied. The resulting Lewis structure for the ammonium ion is:

$$\begin{array}{c} H \\ | \\ H-N-H \\ | \\ H \end{array} \quad +$$

SELECTED REFERENCES

Section 4.1 Molecular Properties

1. T. L. Brown, H. E. LeMay, B. E. Bursten, " Chemistry-The Central Science ", pp 47 - 86, Prentice Hall, 2000
2. M. S. Silberberg, " Chemistry ", pp 61 - 74, Mc Graw Hill, 2000

Section 4.2 Bonding in Molecules

3. T. L. Brown et al., ibid., pp 261 - 289

4 - Molecules

4. M. S. Silberberg, ibid., pp 330 - 353

ADDITIONAL REFERENCES

1. S. E. Manahan, " Fundamentals of Environmental Chemistry ", Lewis, 2001

QUESTIONS

Section 4.1 - Molecular Properties

4.1 Define molecule and molecular formula. Give two examples.

4.2 Distinguish between a molecular formula and an empirical formula. How are they related? Give an example.

4.3 What is meant by molecular mass? What is the molecular mass of each of the following? (a) SO_2 (b) CH_4

4.4 Define the mole. What is the molar mass (in grams) of each of the following? (a) NaCl (b) CO_2

4.5 What is Avogadro's number ? Use it to calculate the number of mols of CH_4 in 3.01×10^{23} molecules of CH_4.

Section 4.2 - Bonding in Molecules

4.6 What is an ionic bond? Illustrate your answer with reference to lithium fluoride, LiF.

4.7 Write the formula of the ionic compound formed between calcium and chlorine. Name three characteristics of ionic compounds.

4.8 Define the covalent bond. Use the Lewis dot structures of the atoms to show the formation of a covalent bond in HBr.

4.9 What is the octet rule? Draw the Lewis structure of CO_2 so that

Questions

each atom obeys the octet rule.

4.10 Write a short essay on the way in which atoms are bonded together in molecules. Illustrate your answer by discussing the bonding in KCl and PH_3.

Chapter 5 - Chemical Change

5.1 Oxidation Numbers

The oxidation number (ON) of an element in a pure substance is a number (positive, zero or negative) which tells us something about the state of combination of the element in the pure substance. These numbers are determined by a set of rules which we now consider:

1. The ON of an element in its free or uncombined state is always zero; for example, $Na_{(s)}$, $Br_{2(l)}$, $N_{2\,(g)}$ all have an ON of 0.
2. The ON of a monatomic ion is equal to the charge on the ion in magnitude and sign; for example, in Na^+Cl^-, ON's of Na and Cl are +1 and -1 respectively. (The magnitude of the charge on the ion is either the group number in the periodic table or 8 - group number for representative elements.
Monatomic ions are formed by the gain or loss of electrons from the neutral atom; for example:

$$Na \rightarrow Na^+ + 1e^-$$
$$Cl + 1e^- \rightarrow Cl^-$$

Other monatomic ions are:
H^+, Li^+, Mg^{2+}, Al^{3+}, Sn^{4+}, Sn^{2+}, S^{2-}, Br^-,
K^+, Ba^{2+}, Pb^{2+}, Pb^{4+}, I^-
Fe^{2+}, Fe^{3+}, Cu^{2+}, Zn^{2+}, Ag^+, Cd^{2+}, Pt^{2+}, Pt^{4+}, Hg^{2+}

3. The ON of certain elements is almost always the same whether in ionic or covalent compounds, e.g F is always -1, alkali metals always +1, oxygen almost always -2, H almost always +1.
4. The algebraic sum of the ON's of all atoms in a polyatomic ion is equal to the overall charge on the ion. Polyatomic ions are made up of two or more elements. They have an overall charge which may be positive or negative.
Common examples are SO_4^{2-}, NO_3^-, CO_3^{2-}, PO_4^{3-}, NH_4^+.
In SO_4^{2-}, the ON of O is -2 and that of S is +6 since $1(+6) + 4(-2) = -2$ (Note that ON refers to one atom of element).
Sometimes we regard the polyatomic ion as a single entity and talk about the oxidation number of the polyatomic ion (which is synonymous with the charge). However, more accurately we

Oxidation Numbers

should consider the oxidation number of each element in the polyatomic ion.

This rule permits us to determine the ON of an element if the ON of every other element in the polyatomic ion is known.

Example:
Determine the ON of N in NO_3^-.
Let x = ON of N
then $1(x) + 3(-2) = -1 \Rightarrow x = +5$
Therefore the ON of N is +5 in NO_3^-.

5. For neutral compounds (ionic or non-ionic) the algebraic sum of the ON's is zero.

What is the ON of N in NO_2?

$1(x) + 2(-2) = 0 \Rightarrow x = +4$
ON of N in NO_2 is +4

(Note that a given element may have more than 1 ON, depending on the compound in which it occurs)

Rules 4 and 5 permit us to write the formulas of new compounds, once we know the ON's of the elements (ions) involved. e.g Aluminum oxide is made up of Al^{3+} and O^{2-}; lowest common denominator of 3 and 2 is 6 $\Rightarrow$ we need 2 Al and 3 O, so the formula of aluminum oxide is Al_2O_3.

Iron (II) sulfate is made up of Fe^{2+} and SO_4^{2-}; its formula is $Fe^{2+}SO_4^{2-}$. Sodium carbonate is Na_2CO_3 and calcium nitrate is $Ca(NO_3)_2$.

Inorganic Nomenclature

Now that we have defined the concept of oxidation number, let us proceed to name systematically inorganic compounds especially binary (two elements) and pseudo binary compounds such as KCN. Some generalizations are:

1. For clearly ionic compounds, the cation is named first and the anion second with the anion ending in 'ide'. For example, Na^+Cl^- is sodium chloride. In order to include covalent compounds, we generalize by naming the less electronegative element first and the more electronegative second with the more electronegative

5 - Chemical Change

ending in "ide". $AlCl_3$ is aluminum chloride.

2. If an element can exhibit more than one oxidation number such as Sn in $SnCl_4$ and $SnCl_2$, the ON of the element is indicated by a Roman numeral in parentheses following the name of the element. For example, $SnCl_2$ is tin (II) chloride and $SnCl_4$ is tin (IV) chloride.

3. For binary covalent compounds, the relative proportions of elements are given by prefixes indicating the number of atoms. SO_2 is sulfur dioxide, Cl_2O_7 is dichloroheptoxide

5.2 Chemical Equations and Stoichiometry

Recall that a chemical change or reaction involves a change in the composition of a substance; new substances called products, are formed by rearrangement of the atoms from the original substances which are called reactants. Very often, it is possible to write a balanced chemical equation for the reaction which shows, in short-hand form, all the reactants and products involved in the reaction in their proper ratios. From the balanced equation, a lot of other quantitative information may be obtained.

We often use the work stoichiometry to describe the combining ratios of elements and compounds; the stoichiometric ratios are the numerical coefficients in the balanced chemical equation. Consider again the combustion of methane, CH_4 (natural gas) to produce CO_2 (a greenhouse gas).

$$CH_{4(g)} + 2\ O_{2(g)} \rightarrow CO_{2(g)} + 2\ H_2O_{(g)}$$

From this equation, we can learn the following:

1. 1 molecule CH_4 reacts with 2 molecules O_2 (or 4 atoms of O) to give 1 molecule CO_2 and 2 molecules of water vapor
2. 1 (6.02×10^{23} molecules) of CH_4 + 2 (6.02×10^{23} molecules) of $O_2 \rightarrow$ 1 (6.02×10^{23} molecules) of CO_2 and 2 (6.02×10^{23} molecules) of H_2O or 1 mol CH_4 + 2 mol $O_2 \rightarrow$ 1 mol CO_2 + 2 mol H_2O (through Avogadro's number)
3. 16 g CH_4 + 64 g $O_2 \rightarrow$ 44 g CO_2 + 36 g H_2O
   Note in (3) that the total number of g on the left = total number of g on the right. This is the Law of Conservation of Mass which says that matter can neither be created nor destroyed.

Types of Reactions

Example:

48 g of CH_4 are burned in a plentiful supply of air. Calculate the mass of CO_2 produced by complete combustion of the CH_4.

$$MM \text{ of } CO_2 = 44, \quad MM \text{ of } CH_4 = 16.0$$

$$48 \text{ g } CH_4 \times \frac{1 \text{ mol } CH_4}{16 \text{ g } CH_4} = 3.0 \text{ mol } CH_4$$

Then $3 \text{ mol } CH_4 \times \frac{1 \text{ mol } CO_2}{1 \text{ mol } CH_4} = 3.0 \text{ mol } CO_2 \times \frac{44 \text{ g } CO_2}{1 \text{ mol } CO_2}$

$$= 132 \text{ g } CO_2$$

5.3 Types of Chemical Reactions

Chemical reactions may be classified as follows:

1. Oxidation-reduction (redox) reaction. An example is:

$$Na_{(s)} + \tfrac{1}{2} Cl_{2(g)} \rightarrow NaCl_{(s)}$$

The overall reaction may be split into two half-reactions:

$$Na \rightarrow Na^+ + 1 \text{ e}^- \text{ (oxidation)}$$
$$Cl + 1 \text{ e}^- \rightarrow Cl^- \text{ (reduction)}$$

Note there is a gain or loss of electrons. Loss of electrons is called an oxidation so that Na undergoes oxidation and is therefore called a reducing agent. Gain of electrons is known as reduction so that Cl undergoes reduction and is thus an oxidizing agent. Also, oxidation involves an increase in the oxidation number of the element oxidized and reduction represents a decrease in the oxidation number of the element reduced.

2. Double Displacement reaction (metathesis). An example is:

$$2 \text{ AgNO}_{3 \text{ (aq)}} + BaCl_{2(aq)} \rightarrow Ba(NO_3)_2(aq) + 2 \text{ AgCl(s)}$$

During a metathesis, there is no change in oxidation number of

5 - Chemical Change

elements.

3. Combination reaction

This is a direct combination of two substances to form a third substance.

e.g. $\quad C_{(s)} + O_{2(g)} \rightarrow CO_{2(g)}$

It may also be a reduction-oxidation or redox reaction as it is in the burning of carbon above.

4. Decomposition reaction

Here, a single substance is converted into two or more substances

e.g. $\quad CaCO_{3(s)} \rightarrow CO_{2(g)} + CaO_{(s)}$

5. Acid/Base (neutralization) reaction

Neutralization involves a reaction between an acid and a base to form a salt and water.

$$HCl_{(aq)} + NaOH_{(aq)} \rightarrow NaCl(aq) + H_2O(l)$$

This is a special case of metathesis reaction since there is no change in oxidation number of any element during neutralization.

Ionic reactions/Equations

Consider again the metathesis reaction:

$$BaCl_{2\,(aq)} + 2\,Ag\,NO_{3(aq)} \rightarrow 2\,AgCl_{(s)} + Ba(NO_3)_{2(aq)}$$

This equation is called the un-ionized, full or molecular equation because each substance is represented by the full chemical formula. Another form of this equation is the total ionic equation, which shows each substance as its ions in solution.

$$Ba^{2+}_{(aq)} + 2Cl^-_{(aq)} + 2Ag^+_{(aq)} + 2NO_3^-_{(aq)} \rightarrow 2\,AgCl_{(s)} + Ba^{2+}_{(aq)} + 2\,NO_3^-_{(aq)}$$

Energy Changes

Notice, however, Ba^{2+} and NO_3^- appear on both sides of the equation, i.e. they have not really taken part in the reaction. They are called spectator ions. If the spectator ions are removed from the equation, we are left with the *net ionic equation*:

$$2\ Ag^+_{(aq)} + 2\ Cl^-_{(aq)} \rightarrow 2\ AgCl_{(s)} \quad \text{or} \quad Ag^+_{(aq)} + Cl^-_{(aq)} \rightarrow AgCl_{(s)}$$

5.4 Energy Changes in Chemical Reactions

Energy

Energy is defined as the capacity to do work or to move matter. Types of energy include kinetic (energy of matter in motion) and potential (energy of matter with potential for motion) A stone perched on top of a hill has potential energy but no kinetic energy. As it moves down the hill, it loses potential energy but gains an equivalent amount of kinetic energy, if we ignore friction. This illustrates two things about energy; first that it is conserved and secondly, that it can be converted from one form to another. Other forms of energy are chemical, electrical, light, and heat. A battery has all four types. In this segment, we are particularly concerned about heat. The units of energy are J (kJ), cal (kcal) and 1 cal = 4.184 J

Heat

What is heat? First, heat is energy. Second, it is energy in transit; it is evident only when it flows and it flows only under the stimulus of a temperature difference. Of course, heat flows spontaneously from a system at higher temperature to a system at a lower temperature and this flow will continue as long as a temperature difference exists. If left in contact long enough, the two systems will finally reach the same temperature and then we say they are in *thermal equilibrium*.

What happens to an object as heat (q) is added to it? There is a rise in temperature (ΔT). How big a rise depends on a quantity called the heat capacity of the substance (C) and the mass (m) of the substance. The relationship is:

5 - Chemical Change

$$q = mC \Delta T$$

or $\quad C = \dfrac{q}{m\Delta T}$

If mass is given in g, then C is called the specific heat of the substance (and written in lower case) and has the units of $JK^{-1}g^{-1}$. Hence the specific heat of a substance is defined as the heat required to raise the temperature of 1 g of the substance by 1 K. If mass is given in mol, then C is called the molar heat capacity (written as upper case) and the units are $JK^{-1}\,mol^{-1}$. For H_2O at 298 K, $c_{H2O} = 1$ cal $K^{-1}g^{-1}$ or 4.184 $JK^{-1}g^{-1}$ and $C_{H2O} = 18$ cal $K^{-1}mol^{-1} = 75.3\;JK^{-1}\,mol^{-1}$.

Heat q is often given a sign: heat absorbed by a substance is considered positive and heat released is negative. Also there are two types of heat capacity depending on whether a process is carried out at constant pressure or constant volume. C_p is the heat capacity at constant pressure and C_v is the heat capacity at constant volume.

Heat changes are measured in an instrument called a calorimeter. For aqueous reactions the calorimeter usually contains a precise amount of water whose temperature rise due to reaction is measured.

Work

We have said that energy is the capacity to do work. But what is work? Work is done when a force (F) applied to an object moves that object a certain distance (ℓ). The magnitude of the work (w) is the product of the force and the distance moved. Like heat, work is often given a sign; it is positive if work is done on the system and negative if work is done by the system.

The First Law of Thermodynamics

Joule arranged experimentally to do work of various kinds on an adiabatic system for which q = 0 (for example, stirring a given amount of water with rotating paddles, or flowing an electric current through a resistance.) He found that for a given amount of work done on the system (water), the rise in temperature of the system (water) was the same regardless of the kind of work done (mechanical, electrical, etc.). This suggested that there was some fundamental property or function (of the

The First Law of Thermodynamics

state) of the system, identified as its internal energy U(or the total energy of the system) that was changing.

The internal energy of a substance may be visualized as the sum of the kinetic and potential energies of the atoms/molecules of the substance. We cannot measure the absolute value of U, but rather, changes in U, as we go from one state to another. The change in internal energy from state 1 to state 2 is found to be equal to the sum of the heat tranferred and the work done on or by the system. Generally:

$$\Delta U = q + w$$

This is a statement of the first law of thermodynamics or the conservation of energy.

We define another heat term called enthalpy or heat content of a substance(H) such that for a process conducted at constant pressure, the change in enthalpy (ΔH) is measured by the heat transferred during the process. For the general reaction:

$$A + B \rightarrow C + D$$

if heat is released in the reaction at constant pressure, then the products together have less enthalpy than the reactants and the difference appears as heat and $\Delta H < 0$. Such a reaction is termed exothermic and an example is

$$H_{2(g)} + \tfrac{1}{2} O_{2(g)} \rightarrow H_2O_{(l)}, \quad \Delta H_{298}^\circ = -285.8 \text{ kJ}$$

Conversely, if heat is absorbed, the reaction is endothermic and $\Delta H > 0$ as in

$$HgO_{(s)} \rightarrow Hg_{(l)} + \tfrac{1}{2} O_{2(g)}, \quad \Delta H_{298}^\circ = 90.0 \text{ kJ}$$

Note that, in addition to specifying the moles of substance(s), we specify the physical state (g,ℓ,s) of the substances involved.

5.5 Rates of Chemical Reactions

Reaction rate

The rate of a reaction is a measure of the speed with which reactants

5 - Chemical Change

are converted into products. It can be defined in terms of the rate of disappearance of reactants or the rate of appearance of products; the two rates are related through the stoichiometric coefficients in the balanced equation for the reaction. Defined in terms of reactants, the rate r_A is defined as the change in number of moles of a reactant A per unit time per unit volume (of the reactor).

Suppose, for the reaction:

$$A + B \rightarrow P \text{ (products)}$$

0.020 mol of A react in a time period of 60 s and the volume of the reactor is 0.50 L, then the rate of the reaction is:

$$\frac{0.020 \text{ mol}}{(0.50 \text{ L})(60 \text{ s})} = 6.7 \times 10^{-4} \text{ mol L}^{-1} \text{ s}^{-1}$$

The units of reaction rate are mol L^{-1} s^{-1}. If we call the number of mol per liter the concentration of a substance, then the rate of a reaction is also the change in concentration of a reactant per unit time. At any given temperature, for example, 298 K, some reactions are naturally fast such as sodium in water:

$$2 \text{ Na}_{(s)} + 2 \text{ H}_2\text{O}_{(\ell)} \rightarrow 2 \text{ NaOH}_{(aq)} + \text{H}_{2(g)}$$

while others are naturally slow such as hydrogen and nitrogen:

$$\text{N}_{2(g)} + 3 \text{ H}_{2(g)} \rightarrow 2 \text{ NH}_{3(g)}$$

Factors Affecting Reaction Rates

Apart from the nature of the reactants, the factors affecting the rate at which a given reaction occurs are:

(a) **Concentration of the reactants** - the rate increases as the concentrations of the reactants increase or

$$r_A = k C_A^a C_B^b$$

Rate of a Reaction

where a is called the order of the reaction with respect to A and b is the order with respect to B; a and b are usually small integers (1,2,3) but may be fractional. a + b is the overall order of the reaction. k is called the rate constant of the reaction at a given temperature and is a measure of the rate of the reaction.

Examples are:

- First order (a = 1); the rate is directly proportional to the concentration of A, that is, $r_A = kC_A$. (Units of k for first order: s^{-1}). The decomposition of dinitrogen pentoxide is an example of a first order reaction.

$$N_2O_5 \rightarrow 2\, NO_{2(g)} + \tfrac{1}{2} O_{2(g)}$$

- Second order; the rate is directly proportional to the product of the concentrations of A and B. Then

$$r_A = kC_AC_B$$

Units of k for second order are: L mol^{-1} s^{-1}. An example is:

$$H_{2(g)} + I_{2(g)} \rightarrow 2\, HI_{(g)}$$

(b) **Temperature** - the rate increases with an increase in temperature and approximately doubles for every 10° C rise in temperature. This behavior can be understood in terms of a theory of reaction rates called the collision theory. This theory says that the reaction rate depends on the frequency of collision of two molecules as well as on their minimum combined energy called the activation energy, E_A. An increase in temperature not only increases their collision frequency, but also increases the number of molecules having the required activation energy, thereby increasing the rate.

(c) **Catalysts** - the reaction rate increases when small amounts of certain substances called catalysts are added to the reaction mixture. Some properties of catalysts are:

(1) they function by reacting chemically with one or more

5 - Chemical Change

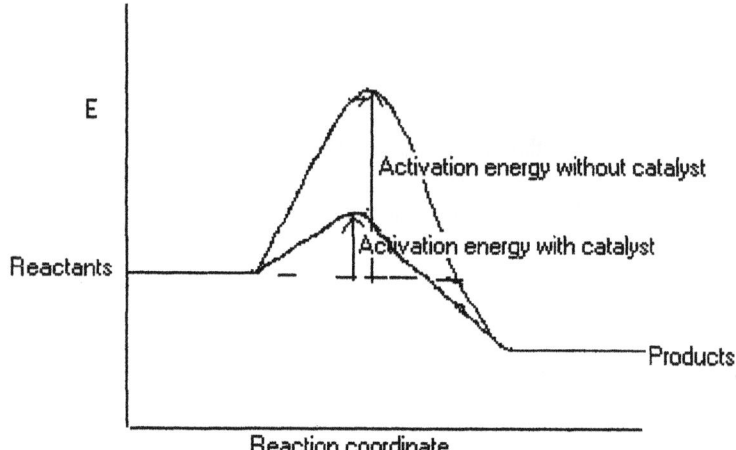

Figure 5.1 The activation energy of a chemical reaction

reactants but is regenerated at the end of the reaction
(2) they increase the rate by lowering the activation energy of the reaction as illustrated in Figure 5.1.

e.g. $H_2O_{2(aq)}$ $\underline{Fe^{3+}}$ $H_2O_{(\ell)} + \frac{1}{2} O_{2(g)}$ (homogeneous catalysis)

$H_{2(g)} + \frac{1}{2} O_{2(g)}$ $\underline{Pt}$ $H_2O_{(\ell)}$ (heterogeneous catalysis)

An intermediate situation occurs in the case of enzymes which are usually proteins catalyzing biological processes. They act as homogeneous catalysts being water soluble. However, they have properties similar to heterogeneous catalysts by orienting the substrates on to their surfaces.

Reaction Mechanisms

A reaction mechanism is made up of a number of **elementary** steps, which, when added together, give the overall reaction. For example, $H_{2(g)}$ and $I_{2(g)}$ react according to the overall equation:

Mechanism of a Reaction

$$H_{2(g)} + I_{2(g)} \rightarrow 2\,HI_{(g)}$$

A mechanism postulated by Sullivan in 1969 has the following elementary steps:

$$I_{2(g)} \rightarrow 2\,I_{(g)}$$
$$H_{2(g)} + 2\,I_{(g)} \rightarrow 2\,HI_{(g)}$$
Overall $\quad H_{2(g)} + I_{2(g)} \rightarrow 2\,HI_{(g)}$

Note that different mechanisms may be suggested for the same overall reaction, but they must all satisfy the criterion that the rate equation derived from the suggested mechanism must agree with the measured rate equation. Mechanisms are inferred from measured reaction rates.

5.6 Chemical Equilibrium

Concept of Equilibrium

Consider the reaction of 3 mols of H_2 and 1 mol of N_2 at 298 K according to the equation:

$$N_{2(g)} + 3\,H_{2(g)} \rightleftharpoons 2\,NH_{3(g)}$$

According to the stoichiometry, we can predict that 2 mol NH_3 will form if all reactants are used up. In reality, the reaction will stop short of exhausting all of the reactants and when no more reaction occurs, the mixture will contain NH_3, N_2, and H_2, the relative amounts depending on the temperature and pressure. Likewise, if we started out with 2 mol of pure NH_3, some of it will decompose to form N_2 and H_2 and we again end up with a mixture of all 3 gases.

In each case, the system is said to be finally in dynamic equilibrium as illustrated in Figure 5.2. The equilibrium is dynamic because both the forward and backward reactions are still occurring at equilibrium, but at the same rate so that there is no net reaction.

The extent of the forward reaction is measured by a quantity called the equilibrium constant for the reaction.

Definition of Equilibrium Constant

For the NH_3 formation reaction as written above, we can define a

5 - Chemical Change

constant called the equilibrium constant K_c in terms of the <u>equilibrium</u> concentrations of the gases in the final mixture, as follows:

$$K_c = \frac{(C_{NH_3}/C°)^2}{(C_{N_2}/C°)(C_{H_2}/C°)^3} = \frac{[NH_3]^2}{[N_2][H_2]^3} = 4.0 \times 10^8 \text{ at } 298 \text{ K}$$

where $[NH_3]$ represents the unitless concentration of NH_3.

For the general reaction:

$$aA + bB \rightleftharpoons cC + dD$$

$$K_c = [C]^c[D]^d/[A]^a[B]^b$$

Qualitatively, K_c is a measure of the position of equilibrium, that is, how far the reaction as written, has proceeded to the right. A large K_c signifies much product and little reactant(s) at equilibrium, whereas a small K_c indicates the reaction as written has hardly occurred.

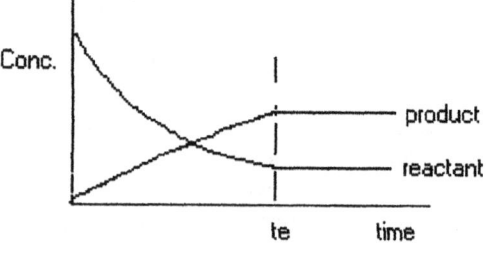

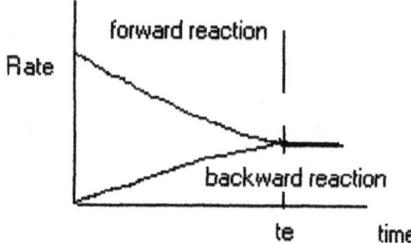

Figure 5.2 The concept of chemical equilibrium

Chemical Equilibrium

Factors Affecting Equilibrium Concentrations

If a system at equilibrium is perturbed in any way, it will readjust itself in such a way as to nullify the perturbation. This is a statement of LeChatelier's Principle which predicts qualitatively the effects of various factors on the position of equilibrium.

The factors affecting equilibrium include:

(a) Concentration of Reactants/Products

For the ammonia reaction, an increase in N_2 concentration causes the equilibrium to shift to the right, resulting in the formation of more NH_3, as predicted by Le Chatelier's Principle. There is no change in K_c of the reaction.

(b) Pressure

Increase in pressure (at constant volume) causes more NH_3 formation as predicted by Le Chatelier's principle. The reaction occurs with a decrease in pressure because 4 mol of reactant gases are converted to 2 mol of product gas at constant volume. There is no effect of pressure if there is no change in number of mols during reaction as in the formation reaction for NO.

$$N_{2(g)} + O_{2(g)} \rightarrow 2\,NO_{(g)}.$$

Again, pressure has no effect on K_c.

(c) Temperature

An increase in temperature (by adding heat) favors the endothermic reaction according to Le Chatelier's Principle and since the decomposition of NH_3 is endothermic, it is therefore, favored at higher temperatures and less ammonia appears in the new mixture.

Generally, K_c decreases with an increase in temperature

5 - Chemical Change

for exothermic reactions and increases with an increase in temperature for endothermic reactions.

(d) **Catalyst** has no effect on the position of equilibrium; it affects the forward and backward reaction rates equally.

SELECTED REFERENCES

Section 5.1 Oxidation Numbers

1. T. L. Brown, H. E. LeMay, B. E. Bursten, " Chemistry - The Central Science", pp 120 - 122, Prentice Hall, 2000
2. M. S. Silberberg, " Chemistry ", pp 149 - 152, McGraw Hill, 2000

Section 5.2 Chemical Equations and Stoichiometry

3. T. L. Brown et al., ibid., pp 86 - 94
4. M. S. Silberberg, ibid., pp 103 - 115

Section 5.3 Types of Chemical Reactions

5. T. L. Brown et al., ibid., pp 105 - 126
6. M. S. Silberberg, ibid., pp 148 - 164

Section 5.4 Energy Changes in Chemical Reactions

7. T. L. Brown et al., ibid., pp 145 - 170
8. M. S. Silberberg, ibid., pp 223 - 236

Section 5.5 Rates of Chemical Reactions

9. T. L. Brown et al., ibid., pp 509 - 547
10. M. S. Silberberg, ibid., pp 664 - 676

Section 5.6 Chemical Equilibrium

11. T. L. Brown et al., ibid., pp 559 - 584

Questions

12. M. S. Silberberg, ibid., pp 714-718, 736-742

ADDITIONAL REFERENCES

1. S. E. Manahan, " Fundamentals of Environmental Chemistry ", Lewis, 2001

QUESTIONS

Section 5.1 - Oxidation Numbers

5.1 What are the common oxidation numbers of the following elements when they are present in compounds? (a) Na (b) O (c) Cl (d) H.

5.2 What is the oxidation number of the named element in the following species? (a) S in SO_2 (b) Mg in MgO (c) N in N_2 (d) C in CO_3^{2-}.

5.3 The cadmium ion is Cd^{2+} and the phosphate and nitrate ions are PO_4^{3-} and NO_3^-. Write the formulas for cadmium phosphate and cadmium nitrate.

5.4 Name the following compounds. (a) $CuSO_4$ (b) Fe_2O_3 (c) $SnCl_4$ (d) SO_3.

Section 5.2 - Chemical Equations and Stoichiometry

5.5 The burning of propane gas, C_3H_8, in oxygen produces carbon dioxide, CO_2, and water. Write a balanced equation for the reaction. How many mols of CO_2 are produced by burning 2 mols of propane?

5.6 Calcium carbonate reacts with hydrochloric acid according to the balanced equation:

$$CaCO_3 \text{ (s)} + 2 \text{ HCl (aq)} \rightarrow CaCl_2 \text{ (aq)} + CO_2 \text{ (g)} + H_2O \text{ (l)}$$

In an experiment, 22.0 g of CO_2 were generated by the reaction. How many grams of calcium carbonate were used up in the reaction?

Section 5.3 - Types of Chemical Reaction

5.7 Define the terms oxidation, reduction and metathesis. Give an example of each.

5.8 Classify the following reactions as redox or metathesis. For the redox reactions, say which element is being oxidized and which is being reduced.

(a) $Fe\ (s) + 2\ H^+\ (aq) \rightarrow Fe^{2+}\ (aq) + H_2\ (g)$
(b) $Cu(OH)_2\ (s) + H_2SO_4\ (aq) \rightarrow CuSO_4\ (aq) + 2\ H_2O\ (l)$
(c) $4\ NH_3\ (g) + 5\ O_2\ (g) \rightarrow 4\ NO\ (g) + 6\ H_2O\ (g)$
(d) $2\ Mg\ (s) + O_2\ (g) \rightarrow 2\ MgO\ (s)$

5.9 Write balanced molecular, total ionic and net ionic equations for the reaction of sodium sulfate with barium chloride in aqueous solution to form a precipitate of barium sulfate and sodium chloride solution.

Section 5.4 - Energy Changes in Chemical Reactions

5.10 Define the terms, energy and heat. Under what conditions would heat flow from one object to another?

5.11 What does the symbol ΔH stand for, with reference to a chemical reaction? If a reaction occurs with an absorption of heat, does the overall enthalpy of the whole system increase or decrease? What is the sign of ΔH?

5.12 What is an exothermic reaction? Is the following reaction exothermic?

$$2\ Mg\ (s) + O_2\ (g) \rightarrow 2\ MgO\ (s), \quad \Delta H = -1204\ kJ$$

Section 5.5 - Rates of Chemical Reactions

5.13 What is meant by the rate of a chemical reaction? What is the effect of the concentration of reactants on reaction rate?

5.14 How does temperature affect the rate of a chemical reaction?

Questions

Explain the effect in terms of collision theory.

5.15 What is a catalyst? Describe two properties of a catalyst.

5.16 What is a *reaction mechanism*? The following mechanism has been proposed for the formation of ozone in the stratosphere. Deduce the overall reaction for ozone formation.

$$O_2 (g) \rightarrow 2 O (g)$$
$$2 O (g) + 2 O_2 (g) \rightarrow 2 O_3 (g)$$

Section 5.6 - Chemical Equilibrium

5.17 Explain dynamic equilibrium as it applies to a chemical reaction. Illustrate with an example.

5.18 A chemical reaction at equilibrium is characterized by an equilibrium constant, K_c. Describe what qualitative information is obtained from a knowledge of K_c.

5.19 State Le Chatelier's Principle. Use it to predict the effect of raising the temperature of an endothermic reaction at equilibrium.

5.20 Consider the following reaction for the formation of NO.

$$N_2 (g) + O_2 (g) \rightleftharpoons 2 NO (g)$$

What is the effect on the equilibrium of (a) increasing the concentration of O_2 (b) removing NO from the mixture as it is formed..

PART B - AIR POLLUTION

Chapter 6 - The Gaseous State

6.1 Properties of Gases

Kinetic Theory of Gases

We have mentioned that in a gas, the forces between molecules are very weak compared to those in liquids and solids. Moreover, the ideal gas is visualized as one in which intermolecular forces are assumed to be entirely absent. Most gases behave ideally (or close to ideal) at very low pressures and relatively high temperatures. This assumption of the absence of intermolecular forces in an ideal gas is part of a theory of the nature of gases called the kinetic molecular theory of gases.

The postulates of the kinetic theory of gases are:

1. A gas is composed of molecules whose size is very small compared to the average distance between the molecules. A consequence of this is that the volume occupied by a gas is mostly free space. (Not so for a liquid or solid). The ideal gas is assumed to move in totally free space.

2. The molecules have a distribution of kinetic energies ($KE = \frac{1}{2} mv^2$) and the average kinetic energy is proportional to the (absolute) temperature; the higher the temperature, the higher the average kinetic energy. This is shown schematically in Figure 6.1.

3. The intermolecular forces in a gas are very weak. In the ideal gas, as we have seen before, these forces are even absent and hence an ideal gas cannot be liquified (by definition).

4. The gas molecules are free to move in all directions, i.e. the motion is random. The pressure of gas is then due to the bombardment of the walls of the container by these molecules.

One of the most important successes of the kinetic theory of gases was that it led to an equation for the ideal gas which agreed well with the empirical ideal gas equation (to be described later). We will omit the details of such developments and go on to talk briefly about the behavior of gases.

Pressure of a Gas

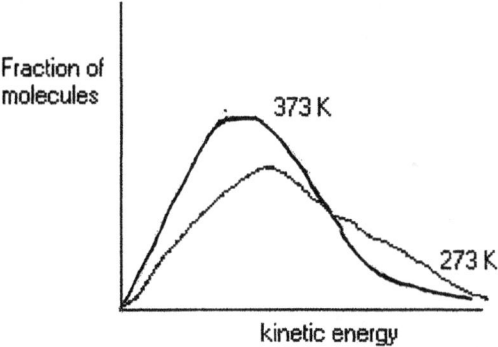

Figure 6.1 Temperature dependence of the distribution of kinetic energies of a gas

Pressure of a Gas

Pressure is defined as the force exerted by a gas on unit area of the surface.

$$P = \frac{F}{A}$$

The unit of pressure in SI units is Nm^{-2} or Pa (for pascal).

The pressure due to the atmosphere may be measured by a mercury (Hg) barometer in which a column of Hg rises into an evacuated tube. This barometric pressure varies from day to day but is approximately 760 mm Hg or 29.92 in Hg. Note, however, that 760mm Hg is defined as the standard atmosphere and must be differentiated from atmospheric (barometric) pressure.

The standard atmosphere is expressed in the following equivalent ways:

1 atm = 14.7 lb in^{-2} absolute (psia) = 29.92 in Hg = 760 mm Hg = 760 torr
= 1.01325 x10^5 Pa = 101.325 kPa

6 - The Gaseous State

The Gas Laws

Consider an ideal gas in which, by definition, intermolecular attractions are absent..

1. For a fixed number of moles (n) of gas at a constant temperature (T), the volume (V) is inversely proportional to the pressure (P). In other words, as the pressure of the gas increases, the volume decreases.
Mathematically:

$PV = k$ (constant)

This is a statement of Boyle's Law.

2. If n and P are constant, then the volume of the gas is directly proportional to the absolute temperature so that as the temperature increases, the volume of the gas increases.
Mathematically:

$V = k'T$

This is Charles' Law.

3. If T and P are constant, the volume is directly proportional to the number of moles of gas.
Mathematically,

$V = k''n$

This is sometimes referred to as Avogadro's Law.

Combining Boyle's and Charles' laws yields the ideal gas equation:

$$PV = nRT$$

in which R is the gas constant and has the value 8.315 $JK^{-1}mol^{-1}$. The ideal gas equation is called an equation of state because it describes mathematically the relationship between the physical properties of the gas.

Ideal Gas Equation

Solids and liquids are easily weighed and also in the case of liquids, volumes are easily measured and masses calculated from densities by the equation:

$$d = \frac{m}{V} \text{ or } m = dV$$

However, for gases we often measure the temperature, pressure, and volume and calculate masses using equations of state of which the ideal gas equation is the simplest:

$$PV = nRT$$

$$n = \frac{PV}{RT}$$

T in this equation must be expressed in K degrees.

The gas constant R has typical units:

$$R = 8.21 \times 10^{-2} \text{ L atm mol}^{-1} \text{ K}^{-1}$$
$$= 8.315 \text{ J mol}^{-1} \text{ K}^{-1}$$
$$= 1.987 \text{ cal mol}^{-1} \text{ K}^{-1}$$

EXAMPLE

A sample of NH_3 gas is contained in a 0.5 L flask at 300 K and 1 atm. Calculate the number of grams of gas present.

$$n = \frac{PV}{RT} = \frac{(1 \text{ atm})(0.5 \text{ L}) \text{ mol K}}{(0.082 \text{ L atm})(300 \text{ K})}$$
$$= 0.020 \text{ mol}$$
$$= 0.020 \text{ mol} \times \frac{17.0 \text{ g}}{1 \text{ mol}}$$
$$= 0.34 \text{ g } NH_3$$

The volume of gas occupied by 1 mol at 273 K and 1 atm (called standard temperature and pressure or STP) can be calculated from the same equation above and is shown to be 22.4 L. This is often called the molar volume at STP and is the same for all **ideal** gases.

6 - The Gaseous State

Dalton's Law of Partial Pressures

Up to now, we have been considering pure gases in using the ideal gas equation. However, we often have to deal with mixtures of gases. For example, air is a mixture of N_2, O_2 and to a lesser extent CO_2. How do we handle these gaseous mixtures?

As you would imagine, in a mixture of different gases, the molecules of each gas are distributed throughout the entire volume, behave more or less independently and contribute, by their impacts on the wall of the container, to the total pressure (P)exerted by the entire mixture. First, this mixture obeys the ideal gas equation ($PV = nRT$), where n is the total number of moles of gas and V is the total volume. Secondly, let us define a key term called the partial pressure (P_i) of component i as that pressure which would be exerted by the given mass (n_i) of the component i if it alone occupied the given total volume V at the given temperature T of the mixture.

Assuming ideal behavior, we can write

$$P_i = \frac{n_i RT}{V}$$

Consider a binary mixture, that is, a mixture of two gases. For each gas, we may write:

$$P_1 = \frac{n_1 RT}{V}, \quad P_2 = \frac{n_2 RT}{V}.$$

It can be shown that:

$$P = P_1 + P_2$$

In other words, the total pressure exerted by a mixture of gases is equal to the sum of their partial pressures. This is one statement of Dalton's Law of Partial Pressures

Real gases

We noted earlier that at relatively high temperatures and low pressures,

Composition of the Atmosphere

most gases behave like ideal gases. However, at low temperatures and high pressures, the molecules of a real gas are fairly close together and intermolecular forces come into play. Also, the actual volume occupied by the gas molecules is no longer negligible. Under these conditions, the ideal gas equation is no longer applicable and we have to find other equations that more nearly represent the behavior of real gases. One of the earliest and simplest was the Van der Waals equation:

$$(P + \frac{n^2 a}{V^2})(V - nb) = nRT$$

where a,b are called Van der Waals constants and depend on the nature of the gas and also on the temperature and pressure. a is a measure of the intermolecular forces (London forces and dipolar forces) in the gas and b is a measure of the molecular or excluded volume.

6.2 Composition of the Atmosphere

Unpolluted air at sea-level consists essentially of N_2 (78% by volume) and O_2 (approximately 20%), with lesser amounts of H_2O, CO_2, Ar and very small amounts of CH_4, NH_3, H_2S, CO, SO_2.

Gases emitted into the atmosphere from natural sources are shown in the following table:

Table 6.1 Concentrations of Trace Atmospheric Gases

GAS	ORIGIN	CONCENTRATION (ppb)
CH_4	Anaerobic biological decay	<65
NH_3	Anaerobic biological decay	
H_2S	Anaerobic biological decay, Volcanoes	
$CH_3 X$ (X=Cl, Br,I)	Oceans	
CO	Fires	<200
SO_2	Volcanoes	
NO_x	Lightning	<10
O_3		<50

Our atmosphere is divided broadly into:

6 - The Gaseous State

(a) the lower atmosphere made up of the troposphere (0-15 km) and the stratosphere (15-50 km). The conditions in the troposphere average to a pressure of approximately 1 atm and 15°C (288 K) at the earth's surface and at the lower limit of the stratosphere to about 10^{-2} atm and -50 °C (223 K) so that the air becomes rarified and the temperature decreases with altitude. At the leading edge of the stratosphere, the temperature begins to rise again due to the absorption of the sun's UV radiation by the ozone layer and reaches about - 2 °C at the upper limit

(b) the upper atmosphere, which consists of the mesosphere (about 50 - 85 km) and the thermosphere (above 85 km). This is shown in Figure 6.3.

The total mass of the atmosphere is estimated to be about 5×10^{18} kg. Because of gravity, most of it is concentrated near the surface of the earth. Let us now look at the individual constituents of air and some of their basic chemistry.

Nitrogen

The symbol is N, the atomic number is 7 and the ground electronic configuration of the atom is $1s^2\ 2s^2 2p^3$. Two isotopes exist: ^{14}N (99.6%) and ^{15}N (0.38%). N has many oxidation states, for example, -3 in NH_3, +2 in NO, +4 in NO_2.

The molecule is diatomic with the two atoms being held together by a triple covalent bond. The Lewis structure of N_2 is

$$:N{\equiv}N:$$

N_2 is a colorless, odorless gas at room temperature and condenses to a colorless liquid at 77 K (-196 °C). Its dissociation energy is 944 kJ/mol which means that it takes that much energy to split the gaseous molecule into gaseous atoms.

$$N_{2(g)} \rightarrow 2N_{(g)}$$

The triple bond structure and large dissociation energy suggest that N_2 is a stable, unreactive molecule. Indeed, at room temperature it hardly reacts, the exception being its reactivity with lithium metal.

$$6\ Li\ (s) +\ N_2\ (g) \rightarrow 2\ Li_3N\ (s)$$

However, it undergoes a number of reactions (the nitrogen cycle) in which

The Nitrogen Cycle

it is continuously being consumed (in sinks)and regenerated (in sources) among the varoius compartments of the earth.

The Nitrogen Cycle

The following are some sources of nitrogen.:

1. $NaNO_3$ (sodium nitrate) is found in oceans and on land as Chile salt peter and $(NH_4)_2SO_4$ (ammonium sulfate) is found on land as fertilizer; these go into making proteins and nucleic acids in organisms. Decay of organic matter by denitrifying bacteria returns nitrogen to the atmosphere. The sequence is $NH_3 \to NO_3^- \to N_2$ (and some N_2O)
2. Liquefaction of air to produce N_2 liquid which is used as a refrigerant.

The following are some sinks for nitrogen:

1. Biological fixation: Nitrifying bacteria take nitrogen from the air and

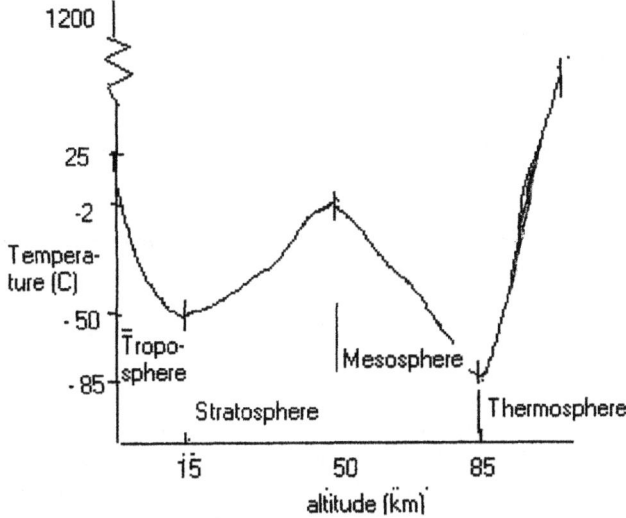

Figure 6.2 The layers of the atmosphere

6 - The Gaseous State

make compounds they need, including NH_3 and NO_3^-

2. Atmospheric fixation, for instance, NO formation during thunderstorms/lightning:

$$N_{2(g)} + O_{2(g)} \rightarrow 2\,NO_{(g)}$$

Then NO gets converted to HNO_3 or NO_3^-.

3. Industrial fixation, for example, NH_3 manufactured by Haber process:

$$N_{2(g)} + 3H_{2(g)} \rightarrow 2\,NH_{3(g)}$$

Then NH_3 is converted to $(NH_4)_2SO_4$ for use in fertilizers.

The net result is that the atmosphere contains about 3.9×10^{18} kg of nitrogen. The residence time of N_2 in the atmosphere is about 10^7 yr. (The residence time is the average time a molecule of the substance spends in a particular sphere of the environment). Hence, there is no danger of the atmosphere being depleted of nitrogen. Moreover, the mixing of nitrogen with other air components is complete thus leading to a uniform distribution of N_2 globally.

Oxygen

The symbol is O, the atomic number is 8 and the ground electronic configuration is $1s^2 2s^2 2p^4$. Three isotopes exist, namely, ^{16}O (99.8%), ^{17}O and ^{18}O. The common oxidation state of O is -2 but it is -1 in the peroxides.

The molecule is diatomic with the two atoms being held together by a double covalent bond. The Lewis structure of O_2 is:

$$:O = O:$$
$$\quad\;\; .. \quad ..$$

O_2 is a colorless, odorless gas at room temperature but condenses to a blue liquid at 90 K (-183 °C). Its dissociation energy is 495 kJ/mol.

Oxygen is reasonably reactive in spite of its double bond and high dissociation energy. Oxidation occurs with both metals and non metals. The resulting compounds are called oxides. Examples are calcium oxide, CaO and carbon dioxide, CO_2. They are formed according to the

The Oxygen Cycle

equations:

$$Ca_{(s)} + \tfrac{1}{2} O_{2(g)} \rightarrow CaO_{(s)}$$

$$C_{(s)} + O_{2(g)} \rightarrow CO_{2(g)}$$

Calcium oxide is called a basic oxide because it is the oxide of a metal while carbon dioxide is an acidic oxide or the oxide of a non-metal.

The Oxygen cycle

The following are sources of oxygen:

1. Biological reduction (photosynthesis) by land and sea plants is the main source of oxygen. The equation is:

$$6\ CO_{2(g)} + 6\ H_2O\ (\ell) \xrightarrow{chlorophyll} C_6H_{12}O_6 + 6\ O_{2(g)}$$

It is believed that the earth's atmosphere was originally devoid of oxygen and that it contained reducing gases such as ammonia, methane and hydrogen. Then, photosynthesis by blue-green algae liberated oxygen converting our atmosphere into an oxygen-rich entity.

2. Smaller amounts of O_2 are produced, mainly in the laboratory, by reactions such as:

$$H_2O_{2\ (aq)} \xrightarrow{Fe^{3+}} H_2O\ (l) + \tfrac{1}{2} O_2\ (g)$$

$$2\ KClO_{3}(s) \xrightarrow[MnO_2]{heat} 2\ KCl\ (s) + 3\ O_2\ (g)$$

The major oxygen sinks are:

1. Biological oxidation (respiration) of carbohydrates by animals; this process is the reverse of photosynthesis.

$$(CH_2O)_n\ (s) + n\ O_2\ (g) \rightarrow n\ CO_2\ (g) + n\ H_2O\ (l)$$

2. Decay of organic matter by aerobic bacteria which uses up oxygen.

6 - The Gaseous State

3. Combustion of fossil fuels and oxidation of metals like iron (rusting).
4. Dissolving of oxygen in the oceans which is used by fish and other marine organisms.
5. Incorporation of oxygen into many rocks and minerals such as silicates and limestone.

The total amount of oxygen in the atmosphere is constant at about 1.2×10^{18} kg. The residence time of O_2 in the atmosphere is about 3,000 yr, and, as in the case of nitrogen, there is no possibility of completely depleting the atmosphere of oxygen. Also, its global distribution is uniform due to thorough mixing.

CO_2

CO_2 is a linear, non-polar molecule whose Lewis structure is written as:

$$:O=C=O:$$

The gas is colorless and odorless at room temperature, but is converted into a white solid (dry ice) on compression.. Dry ice sublimes at -77°C (196 K) and 1 atm and is a very useful refrigerant at that temperature when mixed with solvents like acetone.
For a total pressure of 1 atm, the partial pressure of CO_2 in the atmosphere is about 3.6×10^{-4} atm (or about 360 ppm by volume or about 0.036 % by volume). CO_2 is involved in both the oxygen and carbon cycles.

The Carbon Cycle

The following are <u>sources</u> of CO_2:

1. Biological oxidation (respiration).

$$(CH_2O)_n + n\ O_2 \rightarrow n\ CO_2 + n\ H_2O$$

2. Combustion of fossil fuels

$$2\ C_6H_{14}\ (l) + 19\ O_{2(g)} \rightarrow 12\ CO_{2(g)} + 14\ H_2O_{(l)}$$

The Carbon Cycle

$$C_{(s)} + O_{2(g)} \rightarrow CO_{2(g)}$$

3. Biological decay: e.g. fermentation of sugar to produce alcohol
4. Evaporation from oceans and rivers of dissolved CO_2
5. Volcanic eruptions
6. Heating of mineral carbonates such as limestone:

$$CaCO_{3(s)} \rightarrow CaO_{(s)} + CO_{2(g)}$$

The following are <u>sinks</u> for CO_2:

1. Photosynthesis

$$6\ CO_2\ (g) + 6\ H_2O\ (l) \text{-----} C_6H_{12}O_6\ (s) + 6\ O_2\ (g)$$

2. Dissolution in oceans, rivers and lakes with reaction:

$$CO_{2(g)} + H_2O_{(l)} \rightarrow 2\ H^+\ (aq) + CO_3^{2-}\ (aq)$$

3. Incorporation into minerals like limestone, $CaCO_3$

The average amount of CO_2 in the atmosphere is estimated at about 6.2×10^{14} kg. The residence time is about two years so that CO_2 in the atmosphere is reasonably well mixed and more or less constant globally. Small seasonal variations occur during the year with a peak in April and a trough in October.

Argon

The symbol for argon is Ar, its atomic number is 18 and its electronic configuration is $1s^2 2s^2 2p^6 3s^2 3p^6$. It is a noble gas and is very unreactive. Its boiling point is -185.7°C (87.3K) and its melting point is -189.2 °C (83.8K). For a total atmospheric pressure of 1 atm, the partial pressure of argon is 9.3×10^{-3} atm (or about 9,300 ppm or 0.93% by volume)
Argon is often used as a blanketing atmosphere in electric light bulbs and where there is a need for an unreactive environment.

6 - The Gaseous State

H_2O

Unlike CO_2, the H_2O molecule is bent with a bond angle of about 104°. The Lewis structure is:

$$H-\overset{..}{\underset{..}{O}}-H$$

Its boiling point at 1 atm is 100°C, its melting point is 0°C, its heat of vaporization is 40.7 kJ/mol and its heat of fusion is 6.07 kJ/mol.

For a total barometric pressure of 1 atm, the partial pressure of H_2O varies from about 0 to 0.004 atm depending on the **relative humidity**. The relative humidity is defined as the ratio of the prevailing partial pressure of water to the equilibrium vapor pressure of water at the given temperature. For example, the vapor pressure of H_2O at 25°C is 0.0313 atm so that at 25°, if the partial pressure of water on a given day is 0.065 atm, then the relative humidity is 50%.

The Hydrologic Cycle

The following are the sources of water in the atmosphere:

1. Evaporation from oceans, lakes and rivers.
2. Combustion of fossil fuels.

The following are sinks for H_2O:

1. Precipitation in the form of rain and snow.
2. Photosynthesis (biological reduction) on land and ocean.

The average mass of H_2O in the atmosphere is about 1.0×10^{11} kg. The residence time is about 10 days which is not enough time for good mixing so that there is tremendous variation in $H_2O_{(g)}$ content of air from place to place.

SELECTED REFERENCES

Section 6.1 - Properties of Gases

1. T. L. Brown, H. E. LeMay, B. E. Bursten, " Chemistry - The Central Science", pp 353 - 375, Prentice Hall, 2000
2. M. S. Silberberg, " Chemistry ", pp 176 - 191, McGraw Hill, 2000

Section 6.2 - Composition of the Atmosphere

3. J. M. Beard, " Chemistry, Energy and the Environment ", pp 55 - 64, Wuerz, 1995

ADDITIONAL REFERENCES

1. N. J. Bunce, " Introduction to Environmental Chemistry ", Wuerz, 1993
2. S. E. Manahan, " Fundamentals of Environmental Chemistry ", Lewis, 2001

QUESTIONS

Section 6.1 - Properties of Gases

6.1 State two postulates of the kinetic molecular theory of gases. How are they related to the concept of the ideal gas?

6.2 Define the pressure of a gas. How does the kinetic theory of gases explain the pressure exerted by a gas? What are the units of pressure in SI units?

6.3 State the laws governing the behavior of gases.

6.4 A sample of ammonia gas at 298 K exerts a pressure of 1 atm and occupies a volume of 5.0 L. How many moles of ammonia are present in the sample assuming ideal gas behavior?

6.5 What is meant by the partial pressure of a gas in a mixture? State

6 - The Gaseous State

Dalton's Law of Partial Pressures.

Section 6.2 - Composition of the Atmosphere

6.6 Describe the main layers into which our atmosphere can be divided and say how they differ from each other.

6.7 Name the four main gases present in the atmosphere and give their approximate concentrations as a percentage.

6.8 Write a short essay on nitrogen and the nitrogen cycle.

6.9 Write a short essay on oxygen and the oxygen cycle.

6.10 Write a short essay on carbon dioxide and the carbon cycle.

Chapter 7 - Air Pollution

7.1 Polluted Air

Let us first ask the question, what is polluted air? Polluted air is hazy, cough-producing, eye-watering, generally unhealthy air due to the presence of enhanced levels of gases already present in unpolluted air or to new gaseous contaminants. The substances in polluted air which produce these effects occur in trace amounts, usually at ppb and ppm levels, in unpolluted air. We will discuss the types of contaminants or pollutants, how they arise, i.e. their sources, the kinds of chemical reactions they undergo, and some of the environmental problems they produce such as photochemical smog, acid rain, global warming, and stratospheric ozone depletion. We will also discuss what efforts are being made to improve air quality and to provide a healthier environment for human kind.

7.2 Photochemical Smog

The term smog was originally used to connote the state of polluted air during the industrial revolution and apparently comes from the two words: smoke and fog. It had its origins in the burning of coal during which small particles of unburnt coal (called soot) entered the atmosphere affecting visibility and causing respiratory problems. The combustion of coal produced mostly CO_2 but also some CO.

$$C_{(s)} + O_{2(g)} \rightarrow CO_{2(g)} \text{ (plentiful supply of air)}$$
$$2C_{(s)} + O_{2(g)} \rightarrow CO_{(g)} \text{ (limited supply of air)}$$

Photochemical smog refers to (smog-like) conditions (yellow-brown haze, reduced visibility, eye-watering, nose-itching) initiated by sunlight and occurs classically in big cities like Los Angeles with lots of automobiles.

There are five conditions necessary for the formation of photochemical smog:

1. Nitrogen oxides, NO_x (NO/NO_2) in the air
2. Hydrocarbons, such as CH_4, gasoline and unsaturated hydrocarbons
3. Sunlight
4. Temperatures above 18°C
5. Low wind and inversion of temperature

7 - Air Pollution

If all these conditions are satisfied, the end result is the addition to the atmosphere of increased levels of NO_2, O_3, CO and an organic compound called peroxyacetyl nitrate (PAN), all of which are very injurious to human health. Let us now see how the levels of these dangerous gases are enhanced, because as mentioned earlier, these compounds are present in trace amounts even in unpolluted air.

Table 7.1 Concentrations of Air Pollutants

Pollutant	Unpolluted air (ppb)	Polluted air (ppb)
CO	<200	10^4- 5 x10^4
NO_x	<10	10^3- 3.0 x10^3
O_3	<50	100-500
Hydrocarbons	<65	1.5x10^3
PAN	<0.05	20-70

The process begins with the incomplete combustion of gasoline and the combustion of N_2 at the high temperatures of the engine to form three <u>primary</u> pollutants, NO, CO and unreacted hydrocarbons. The reactions are:

$$2\ C_6H_{14\ (l)} + 19\ O_{2(g)} \rightarrow 12\ CO_{2(g)} + 14\ H_2O_{(l)} \text{ (complete combustion)}$$
(Hexane)

$$2\ C_6H_{14(l)} + 13\ O_{2(g)} \rightarrow 12\ CO_{(g)} + 14\ H_2O_{(l)} \text{ (incomplete combustion)}$$

$$N_{2(g)} + O_{2(g)} \rightarrow 2\ NO_{(g)} \text{ (endothermic)}$$

NOTE: NO is a useful chemical in the human body. It helps to keep arteries open and so helps to reduce strokes and heart attacks. The 1998 Nobel Prize for Medicine was awarded to the people who worked out how NO prevents strokes, etc.

Nitric oxide (NO) is an oxide of nitrogen (an oxide is a compound of oxygen and another element). The Lewis structure of NO is:

$$:\!N = O\!:$$

Photochemical Smog

and contains an odd electron. It is a colorless gas but condenses to a blue liquid. The formation of NO from N_2 and O_2 is endothermic and at room temperature the equilibrium lies far to the left ($Kp_{298} = 4.1 \times 10^{-31}$), that is, very little NO exist in the equilibrium mixture at 298 K. Also the reaction rate is slow because of the large activation energy. However, when the temperature is high as in the automobile engine, the equilibrium shifts to the right according to Le Chatelier's principle ($Kp_{2400} = 5.0 \times 10^{-2}$).

Furthermore, as the NO exits in the automobile exhaust, it cools but the reaction does not shift back to the left because of the slow rate of the reverse reaction, again due to the high activation energy of the reverse reaction.

Next, NO is oxidized to NO_2 but not directly by O_2 (as you might suspect) because the reaction:

$$NO_{(g)} + O_{2(g)} \rightarrow 2\, NO_{2(g)}$$

is very slow at room temperature. Instead, certain (unsaturated) hydrocarbons act as intermediates for oxidation of NO by O_2.

$$NO\,(g) + ROO. \rightarrow NO_2\,(g) + RO.$$

Also, powerful sinks for NO are ozone, O_3 (which is a very strong oxidizing agent) and the the hydroperoxyl radical, .OOH.

$$NO_{(g)} + O_{3(g)} \rightarrow NO_{2(g)} + O_{2(g)}$$

$$NO(g) + HO_2.\,(g) \rightarrow NO_2(g) + HO.$$

The NO_2 formed is called a <u>secondary</u> pollutant. It is toxic and has an unpleasant odor. NO_2 is a red-brown gas. Its Lewis structure is

$$:\!\overset{..}{O}\!::\!\overset{..}{N}\!:\!\overset{.}{O}\!:$$

However the molecule is bent (bond angle of 134°) and contains an odd electron as in NO. NO_2 is a reactive molecule and undergoes many reactions, two of which are relevant to our discussion.

1. Formation of HNO_3 (acid rain)

$$4\, NO_{2(g)} + 2\, H_2O_{(\ell)} + O_{2(g)} \rightarrow 4\, HNO_{3(\ell)}$$

7 - Air Pollution

2. Photochemical decomposition. Its brown color suggests that NO_2 absorbs sunlight both in the visible and UV part of the electromagnetic spectrum. However, only UV radiation ($\lambda < 400$ nm) can cause decomposition of NO_2 to reform NO.

$$NO_{2(g)} \xrightarrow{\lambda \leq 400 \text{ nm}} NO_{(g)} + O_{(g)}$$

The $O_{(g)}$ atom, a very reactive species, is also formed. It can undergo a number of reactions, one of which is the formation of ozone from oxygen:

$$O_{2(g)} + O_{(g)} \rightarrow O_{3(g)}$$

The above reaction can be reversed photochemically by UV radiation of a shorter wavelength (and greater energy)

$$O_3 \xrightarrow{\lambda \leq 320 \text{ nm}} O_2 + O^*$$

O^* represents a highly energetic (excited) O atom which is capable of reacting further to form, among other things, the .OH radical.

$$O^* + H_2O \rightarrow 2 \cdot OH_{(g)}$$

The • OH radical is the major player in tropospheric chemistry (polluted or not). For example, it is involved in the oxidation of CO to CO_2.

$$\cdot OH_{(g)} + CO_{(g)} \rightarrow H + CO_2$$

(CO is a very hazardous gas because it combines with hemoglobin to form carboxy hemoglobin making the haemoglobin unavailable for transporting O_2 to cells)

Another reaction of •OH is hydrogen abstraction from hydrocarbons, e.g. CH_4 to form ultimately HCHO and O_3.

Overall equation: $$CH_{4(g)} + 4 O_{2(g)} \xrightarrow{h\nu} HCHO_{(g)} + H_2O + 2 O_{3(g)}$$
(formaldehyde)

A third reaction of •OH is reaction with NO_2 to form HNO_3 (nitric acid)

Photochemical Smog

Overall equation: $4 NO_{2(g)} + 2 H_2O_{(l)} + O_{2(g)} \rightarrow 4 HNO_{3(l)}$

Peroxyacetyl-nitrate (PAN - $CH_3CO \cdot OONO_2$) is formed in a side reaction in the oxidation of hydrocarbons through the intermediary of compounds called aldehydes (formaldehyde above is an example).

$$C_6H_{14} \underline{} O_3 \text{aldehydes} \underline{\text{(sunlight)}} \text{PAN}$$

PAN is a lachrymatory (causes tearing). It is very toxic to both plants and animals

The above processes for the formation of secondary pollutants, (NO_2, O_3, HNO_3, PAN) are aided by a temperature in excess of 18°C at which they proceed at reasonably fast rates.

Secondary pollutants such as ozone and PAN are toxic not only to animals and humans but also to plants. They inhibit photosynthesis thus causing large financial losses in agriculture.

Control of Photochemical Smog

Automobile exhaust is the major source of NO_x and hydrocarbons in the atmosphere. As we have seen, these pollutants are partly responsible for photochemical smog, so that a strategy to reduce photochemical smog must begin with elimination or control of these gases in auto exhaust.

1. First and foremost, we must reduce use of automobiles to a minimum, for example, by using public transportation and car-pooling as much as possible.
2. If this is not practical, as in New Jersey, automobile engines must be kept well tuned to reduce emission of unburnt fuel and NO_x.
3. Emission control devices such as catalytic converters must be fitted to all cars. Catalytic converters act by creating a second zone for combustion of spent fuel at a relatively low temperature. The catalysts used are finely divided precious metals like Pt, Pd and Rh supported on an inert matrix made up of the oxides of aluminum, silicon and magnesium. These catalysts are poisoned by lead so that lead-free gasoline is necessary for the proper operation of these catalytic converters.

7 - Air Pollution

The reactions involved in the two-stage converter are as follows:

1. The first stage involves reduction of NO by CO and H_2 in the presence of Rh as catalyst. The CO and H_2 are formed by the steam reforming of hydrocarbon in presence of catalyst.

$$CH_{4(g)} + H_2O_{(g)} \xrightarrow{Rh} CO_{(g)} + 3 H_{2(g)}$$

$$H_{2(g)} + 2 NO_{(g)} \xrightarrow{Ni} N_{2(g)} + 2 H_2O_{(g)}$$

$$2 CO_{(g)} + 2 NO_{(g)} \rightarrow N_{2(g)} + 2 CO_{2(g)}$$

2. The second stage involves oxidation of hydrocarbons and CO in the presence of Pt and Pd as catalysts.

$$2 C_6H_{14} + 19 O_2 \xrightarrow{Pt} 12 CO_2 + 14 H_2O$$

$$2 CO_{(g)} + O_{2(g)} \rightarrow 2 CO_{2(g)}$$

Present converters can eliminate 96% of unburnt hydrocarbons and CO and 76% of NO_x emissions.

7.3 Stratosphere Ozone Depletion

We have seen that the build-up of ozone in the troposphere (ground level ozone) is a serious health problem especially for people with respiratory problems like emphysema. In the stratosphere, however, we need to keep the O_3 concentration up because it plays such a crucial role in protecting humans from the harmful effects of UV radiation. So the depletion of ozone in the stratosphere by certain chemicals is a serious global problem.

Let us begin by reviewing the nature of the stratosphere, its composition and chemistry. Recall that the stratosphere extends approximately from about 15 km to 50 km and that the temperature is actually increasing from about -50°C (223K) at the bottom to about 0° (273K) at the top. (Beyond the stratosphere is the mesosphere where the temperature begins to decrease again). Unlike temperature, the pressure (P)decreases steadily with altitude so that at the bottom of the stratosphere P is about 0.1 atm and at the top, P is about 10^{-4} atm. Even though the air

Stratospheric Ozone

is so rarefied, its composition remains more or less the same as in the troposphere, with N_2 and O_2 still being the major constituents.

Stratospheric ozone occurs at a steady state concentration of about 10 ppm, which seems like an insignificantly small concentration but happens to be very important in the chemistry of the stratosphere. (One small point: we often read in the newspapers about the destruction of the ozone layer. Strictly speaking, there is no layer of O_3 in the stratosphere; O_3 is more or less uniformly distributed through the stratosphere)

Ozone Formation and Destruction

O_3 is formed in the stratosphere in much the same way it is formed in the troposphere (from O_2):

$$3\ O_{2(g)} \rightarrow 2\ O_{3(g)}$$

The mechanism is:

$$O_{2(g)} \xrightarrow{\lambda \leq 240\text{ nm (UV-C)}} 2\ O_{(g)}, \quad \Delta H° = 496 \text{ kJ}$$

$$2\ O_{(g)} + 2\ O_{2(g)} \rightarrow 2\ O_{3(g)}, \quad \Delta H° = 2 \times (-105) = -210 \text{ kJ}$$

Overall $\quad 3\ O_{2(g)} \xrightarrow{\lambda \leq 240 \text{nm}} 2\ O_{3(g)}, \quad \Delta H° = +286 \text{ kJ}$

so that the overall process is endothermic and the energy comes from sunlight. The above reactions are all reversible and account for the destruction of O_3. Ozone absorbs radiation in the range 250-330 nm.

$$O_{3(g)} \xrightarrow{\lambda \leq 325 \text{ nm}} O_{2(g)} + O_{(g)}, \quad \Delta H^0 = 105 \text{ kJ}$$

$$O_{(g)} + O_{3(g)} \rightarrow 2\ O_{2(g)} \quad \Delta H^0 = -389 \text{ kJ}$$

Overall $\quad 2\ O_{3(g)} \rightarrow 3\ O_{2(g)}, \quad \Delta H° = -286 \text{ kJ},$

So normally, ozone is being destroyed at about the same rate as it is being reformed, thus maintaining a steady state concentration of about 10 ppm. However, in the process, UV energy from sunlight is absorbed and converted to heat. This is the major mechanism for the heating of the stratosphere and causes the top of the stratosphere to be warmer than the

7 - Air Pollution

troposphere.

Enhanced Destruction of the Ozone layer

The destruction of the ozone layer only becomes problematical when it is enhanced by the presence of certain catalytic substances in the stratosphere. Recall what a catalyst does. It speeds up a given reaction by lowering the energy barrier or activation energy, is involved in the reaction but is regenerated at the end.
Some of the catalysts found in the stratosphere are the Cl and Br atoms produced by the photolysis of CFC's and halons respectively.

CFC is an acronym for chlorofluorocarbon which contains chlorine, fluorine and carbon. The most important commercially are CFC-11 ($CFCl_3$) and CFC-12 (CF_2Cl_2). CFC-12 was developed in the 1930's to replace the highly toxic gases NH_3 and SO_2 as refrigerants in air conditioners and refrigerators. (In refrigeration, the refrigerant gas is liquified by compression, then as the liquid flows through the coils bordering your food, it is vaporized by absorbing heat from the food; the vapor is then compressed, and the cycle repeated.). NH_3 has a BP of -33°C whereas CFC-12 is -30°C. NH_3 has a $\Delta H_v = 23$ kJ/mol. CFC-12 has $\Delta H_v = 35$ kJ/mol.

CFC-11 was used as a blowing agent in plastic foams and as a propellant in aerosols. Both CFC's were deemed extremely useful because they were non-toxic, non-flammable, unreactive, odorless compounds. 1980 production of CFC's was about 700,000 tons of which about 600,000 tons escaped into the troposphere, then into the stratosphere. Troposphere concentrations of CFC-11 are estimated at about 150 ppt (1979).

With CFC-12, the reactions are:

CF_2Cl_2 $\lambda < 250$ nm $CF_2Cl\cdot + Cl\cdot$ (1) initiation

$O_{3(g)} + Cl\cdot_{(g)} \rightarrow ClO_{(g)} + O_{2(g)}$ (2) propagation

$ClO_{(g)} + O_{(g)} \rightarrow Cl\cdot + O_{2(g)}$ (3) propagation

Overall $O_{3(g)} + O_{(g)}$ $\underline{\cdot Cl}$ $2 O_{2(g)}$

Note that the Cl• reaction can be repeated many times. On average, one Cl atom leads to the destruction of about 20,000 O_3 molecules. Reactions of this type are called chain reactions. The initiation and propagation steps are as shown above. There is also a termination step in which Cl atoms are ultimately destroyed by forming stable chlorine compounds such as chlorine nitrate ($ClNO_3$) and hydrogen chloride (HCl)

$$Cl\bullet + HNO_3 \rightarrow ClNO_3 + H\bullet$$

$$Cl\bullet + H_2O \rightarrow HCl + \bullet OH$$

Halons are partly brominated fluorocarbons used in fire extinguishers, for example, CF_3Br (H-1301) and CF_2ClBr (H-1211). They behave similarly to the non Br-containing CFC's by releasing catalytic quantities of Br atoms by photolysis.

$$CF_2ClBr \rightarrow CF_2Cl + Br$$
$$Br + O_3 \rightarrow BrO + O_2$$
$$BrO + O \rightarrow Br + O_2$$

They are even more effective ozone-depleting substances (ODS) than CFC-11 and CFC-12 because of the weakness of the C-Br bond

CFC's and halons have posed a serious problem for stratospheric ozone because they are long-lived pollutants. They are completely unreactive in the troposphere and react relatively slowly even in the stratosphere. As a result of their ozone depleting potential (ODP), they have been banned under the 1987 Montreal Protocol and their production has being phased out.

The Antarctic Ozone Hole

You may have heard about the 'hole' in the ozone layer over Antarctica which develops in late winter and destroys about 50% of the stratospheric ozone. The process of ozone destruction we have discussed so far, occurs in the presence of sunlight and therefore cannot occur during the long dark Antarctic winter. Therefore, another mechanism has been proposed to account for this phenomenon. Briefly, it has been suggested that Cl_2 and HOCl are produced on the surface of $HNO_3 \bullet 3H_2O$ and ice crystals at low temperatures by the reaction of chlorine nitrate:

$$HCl + ClNO_3 \rightarrow Cl_2 + HNO_3$$

7 - Air Pollution

$$H_2O + ClNO_3 \rightarrow HOCl + HNO_3$$

Cl_2 and HOCl then absorb in the visible and near UV to form Cl atoms again because this can take place at the low light levels of the polar sunrise in late winter/early spring. Thus, destruction of O_3 by Cl atoms occurs without comparable O_3 formation in late winter/early spring. However, as spring progresses, ice sublimes, sunlight levels rise and normal ozone formation and destruction are resumed.

Consequences of Ozone Depletion

We have mentioned that stratospheric ozone is important in that it reduces the amount of harmful UV radiation reaching the surface of the earth. Hence there are seroius consequences for its depletion in the stratosphere. These include:

1. Increased levels of UV-B ($\lambda \geq$ 290 nm) and some UV-C ($\lambda <$ 290 nm) at the earth's surface could adversely affect plants, micro- organisms, phytoplankton, even animal life, thus disrupting the food supply. The biological effect of UV is that it damages the genetic material, deoxyribonucleic acid (DNA) of cells. DNA is an efficient absorber of UV radiation and undergoes photochemical decomposition when it absorbs UV radiation.
2. For humans, increased incidence of skin cancer occurs, especially melanoma, among light - skinned people. Dark-skinned people have a skin pigment called melanin which absorbs UV radiation effectively and reduces the probability of UV reaching living cells under the skin. Light skinned people are therefore well-advised to protect themselves against the negative effect of UV by using sun screens (such as SF 15) when exposed to intense sunlight. Sunscreens are man-made chemicals which act like melanin by absorbing UV and thus preventing it from reaching living cells below the skin.

Strategies for Controlling Ozone Depletion

There are several strategies that can be used to control stratospheric ozone depletion.

Acid Rain

1. Developed countries like the U.S. and Europe have signed agreements such as the Montreal Protocol of 1987 which called for the complete phase out of CFC production by the year 2,000. However, even if production stopped completely today, the problem will still be with us for a long time to come because of the unreactivity and long lifetime (about 100 years)of CFC's in the troposphere and stratosphere.

CFC's are very useful compounds and their banning means that we need replacement compounds that are hopefully as inert, non-toxic and non-flammable as CFC's but which can be degraded completely in the troposphere or the stratosphere. Non-chlorine CFC's that are being developed include CF_3CH_2F (CFC-134a) and CH_3CHF_2 (CFC-152). They both have zero ozone-depleting-potential (ODP). CFC-152 is, however, somewhat flammable but is already available commercially from DuPont. CFC-134a is at the pilot plant stage of development (2000)

2. The measures above are preventive measures which will not get rid of the existing CFC's in the stratosphere. Some scientists have suggested injecting a substance such as propane into the stratosphere to act as a sink for Cl atoms. Thus the propane will compete with ozone for Cl atoms and hence reduce the destruction of ozone by Cl atoms. Theoretically the concept is a good one but its implementation would require a massive injection of propane worldwide and might be impractical. Also, the injection of such large amounts of propane may have unforeseen effects and may, in fact, worsen the situation.

7.4 Acid Rain

Acid rain is a very serious regional environmental problem as distinct from a global problem like stratospheric ozone depletion. It costs millions, if not billions, of dollars in lost resources, both agricultural and marine. So what is acid rain? It is a component of "acid deposition" which is a generic term for the atmospheric precipitation of acids and includes acid fog and acid snow, besides acid rain. (The term "dry deposition" is sometimes used to describe the direct interaction of gases like SO_2 and HCl with plants and animals).

Ordinary rain is slightly acidic (pH of about 5.6) due to the presence of dissolved CO_2. The equation for the reaction is:

$$CO_{2(g)} + H_2O_{(l)} \rightleftharpoons 2\ H^+_{(aq)} + CO_3^{2-}_{(aq)} \quad (H_2CO_3 \text{ or carbonic acid})$$

7 - Air Pollution

So 'acid rain' is rain that is considerably more acidic than this, with a pH less than 5.6.

Origin of Acid Rain

1. We have seen that NO_x is generated in automobiles and power stations. A primary source of acid rain is NO_2 which reacts with water, in the presence of oxygen, to form nitric acid, HNO_3.

$$4\ NO_{2(g)} + 2\ H_2O_{(l)} + O_{2(g)} \rightarrow 4\ HNO_{3(aq)}$$

HNO_3 is a very strong acid (one of the six strongest acids mentioned earlier). Its Lewis structure is:

$$\begin{array}{c} HO-N=O \\ \| \\ O \end{array}$$

Its strength is due to its instantaneous reaction with water to produce a high concentration of hydrogen ions, H^+.

$$HNO_3\ (aq) \rightarrow H^+ + NO_3^-$$

When concentrated, it would burn your skin. However, it occurs at too low a concentration in acid rain to be corrosive. Nevertheless, it still has a devastating effect on plants and marine species. The effect is seen over wide areas because of Long Range Transport of Atmospheric Pollutants (LRTAP); the wind can carry these pollutants hundreds of miles away from their sources. For example, most of the acid rain that falls in Norway is believed to originate elsewhere in Europe.

2. A second source of acid rain is sulfur dioxide, SO_2. Its Lewis structure is:

$$:\ddot{O}-\ddot{S}=\ddot{O}$$

SO_2 is produced naturally in volcanoes and to a lesser extent by plants.

Acid Rain

The concentration in 'clean' air is about 100 ppb (0.1 ppm). However, there are a number of man-made (anthropogenic) sources, chief among them being the burning of high-sulfur coal in power generating stations. Coal contains 1- 5 % sulfur mainly in the form of iron pyrite, FeS_2 and organic sulfur. The oxidation of pyrite during combustion is given by the equation:

$$4\ FeS_{2(s)} + 11\ O_{2(g)} \rightarrow 2\ Fe_2O_{3(g)} + 8\ SO_{2(g)}$$

. Partial removal of the sulfur may sometimes be accomplished by pulverizing the coal and removing the included FeS_2 by washing. The more dense pyrite sinks to the bottom. This process adds to the cost of power generation. Therefore, many power stations dispense with this step and their smoke stacks contain significant amounts of SO_2.

A second source of SO_2 is the smelting of non-ferrous ores. For example, the first step in the extraction of nickel is:

$$2\ NiS_{(s)} + 3\ O_{2(g)} \rightarrow 2\ NiO_{(s)} + 2\ SO_{2(g)}$$

In many cases, the SO_2 is trapped by passing it over calcium oxide, CaO with which it forms calcium sulfite, $CaSO_3$.

$$CaO(s) + SO_2(g) \rightarrow CaSO_3(s)$$

The calcium sulfite is subsequently converted to the valuable by-product, calcium sulfate ($CaSO_4$) by oxidation. However, in some cases, SO_2 is simply emitted into the air and is then a serious pollutant.

A third source of SO_2 is the burning of pure S as the first step in the manufacture of the important chemical, sulfuric acid, H_2SO_4.

$$S(s) + O_2(g) \rightarrow SO_2(g)$$

Once SO_2 gets into the atmosphere, it can dissolve by reaction with H_2O in the presence of oxygen to form sulfuric acid, H_2SO_4.

$$2\ SO_{2(g)} + 2\ H_2O_{(l)} + O_{2(g)} \rightarrow 2\ H_2SO_{4(aq)}$$

H_2SO_4 is, like HNO_3, a very strong acid and leads to ecological devastation when present in acid rain for the same reasons.

7 - Air Pollution

$$H_2SO_{4(aq)} \rightleftharpoons 2H^+ + SO_4^{2-}$$

By all accounts, sulfuric acid is the major component of acid rain.

Effects of Acid Rain

The predominant effects on the environment of acid rain are:

1. Ecological devastation of plants and marine organisms. The devastation depends on the composition of the soil and bedrock in the local area. If the bedrock is limestone or chalk ($CaCO_3$), then it reacts with the acid and effectively eliminates any adverse effects. The reaction is:

$$CaCO_{3(s)} + 2 H^+ \rightarrow Ca^{2+}_{(aq)} + CO_{2(g)} + H_2O_{(\ell)}$$

or specifically with HNO_3 :

$$CaCO_{3(s)} + 2 HNO_{3(aq)} \rightarrow Ca(NO_3)_{2(aq)} + CO_{2(g)} + H_2O$$

However, if the bedrock is granite or quartz, it will not be able to neutralize the acid and devastation results. In these cases, there may be deliberate addition of lime (CaO) to help neutralize the acid.

2. Erosion of structures made of limestone by the same reaction quoted above. Metal statues are also corroded.

$$Fe_{(s)} + H_2SO_{4(aq)} \rightarrow FeSO_{4(aq)} + H_{2(g)}$$

3. Leaching out of aluminum ions (Al^{3+}) from soil and rock due to the acidity. Al^{3+} in higher than usual concentration is found to be detrimental to fish and aquatic plants.

4. Respiratory problems in humans and animals due to dry deposition.

Control Strategies for Acid Rain

1. SO_2 emission standards have been set in the U.S. by the Clean Air Act (1970, 1977, 1990). Companies using coal are required to reduce their SO_2 emissions by incorporating scrubbers (the equivalent of catalytic

Global Warming

converters in cars) in the processes. As a matter of fact, SO_2 allotments are now traded like commodities. Each company is allowed a certain level of SO_2 emission. If it does not use up its allowance, it can sell the excess to another company that exceeded its allotment.

2. SO_2 emissions from power plants can, of course, be reduced by substituting oil and natural gas for coal but that makes the energy more expensive, so as usual, a balance has to be found.

7.5 **Global Warming**

The earth radiates heat constantly even as it receives radiation of varoius kinds from the sun. If the heat from the earth could escape completely into the stratosphere and beyond, the temperature at the surface of the earth would be much less than the average 15°C (288 K) that exists today. Why is the temperature maintained at about 15°C? It is because certain gases in the troposphere, notably CO_2, H_2O, CH_4 and N_2O, absorb some of the heat (radiation) coming from the earth and re-radiate it back, thus preventing it from being all lost into space. Therefore, the surface of the earth is kept relatively warm.

These gases, like CO_2 and H_2O, are known as greenhouse gases and the effect is called the greenhouse effect by analogy with a greenhouse which, while letting in sunlight, prevents heat radiation from leaving the enclosed space. (Strangely enough, the major tropospheric gases, N_2 and O_2, are not involved in the greenhouse effect because they do not absorb heat radiation).

Before we get into the details of the greenhouse effect, (and the enhanced greenhouse effect which gives rise to global warming), let us first talk about the nature of radiation and in particular, heat or infrared radiation.

Electromagnetic radiation (EM) sometimes behaves like a wave. What is a wave? It is a periodic or cyclical variation of some quantity with time or distance in a medium that may be vacuum or air. With water waves, the quantity varying is the number of H_2O molecules. With EM radiation, it is the electric and magnetic fields associated with EM radiation. Waves are characterized by a wavelength (λ) which is the distance from peak to peak or trough to trough of the wave.

Units of λ are the units of distance, namely m, cm (10^{-2} m), μm, (10^{-6} m), nm (10^{-9} m). Visible radiation goes from about $\lambda = 400$ nm (violet) to

7 - Air Pollution

about $\lambda = 800$ nm (red), UV radiation goes from about $\lambda = 400$ nm to about $\lambda = 10$ nm and X-radiation has λ less than 0.1 nm. Heat or infra-red (IR) radiation goes from about $\lambda = 800$ nm (0.8µm) to about $\lambda = 25$ µm. A wave is also characterized by a velocity which is the distance travelled per unit time. The velocity (c) of EM radiation is 3.0×10^8 m s^{-1} (300,000 km/s) which is very fast indeed. Also all radiation, including heat or infrared radiation, sometimes behaves like a particle. These packets of energy are often called photons

Molecules will absorb IR radiation depending on their structure. Certain molecules like N_2 and O_2, which are homonuclear and non-polar, cannot absorb IR radiation (because their vibrational motion does not lead to a change in the polarity of the molecule). The effective IR absorbers in the atmosphere are CO_2, CH_4, H_2O, N_2O and CFC's. When these atmospheric molecules absorb IR radiation coming from the earth's surface, they re-radiate IR in all directions and some of this IR is redirected back to earth. The net result is that the earth's surface is kept at a comfortable 15 °C by these greenhouse gases when they are present at normal concentrations.

Substantially increased concentrations of these greenhouse gases would

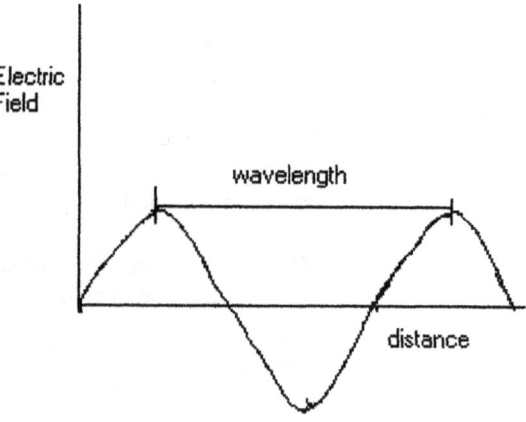

Figure 7.1 The wave properties of electromagnetic radiation

Global Warming

trap more IR than usual, thereby returning more IR to the earth (the enhanced greenhouse effect) and causing an increase in the average temperature of the troposphere. The phenomenon is known as global warming.which could result in an increase in temperature of 1 - 5 °C over the next century. It is being predicted by some scientists that global warming could have catastrophic consequences for the global climate and therefore must be brought under control.

Let us first discuss greenhouse gases and some of their chemistry. We have already discussed CO_2 (which happens to be the main greenhouse gas accounting for about 50% of global warming), H_2O (which contributes about 5%)and CFC's (which contribute 20%) so that we will concentrate here on CH_4 (which contributes 20%) and N_2O (about 5%).

Methane, CH_4

The Lewis structure of methane is:

$$\begin{array}{c} H \\ | \\ H-C-H \\ | \\ H \end{array}$$

It is the simplest hydrocarbon (a compound of hydrogen and carbon only) and belongs to a family of hydrocarbons called the alkanes. The steady state concentration of tropospheric CH_4 is now about 1.7 ppm (up from 0.75 ppm in the 1700's so that it is rising gradually at about 1- 2% per year).

Sources

Methane is generated from the following sources:

1. Biological (anaerobic) decomposition of plant material, for example, in wetlands and rice paddies. This gives rise to the names 'marsh gas'and 'swamp gas' to describe methane.
2. Ruminant animals such as cattle, sheep , buffalo. These animals produce CH_4 in their stomachs as a result of digestion (anaerobic fermentation) of cellulose in their food. Their belching spews CH_4 into

7 - Air Pollution

the air.
3. Anaerobic decomposition of organic matter (in garbage) in landfills.
4. Biomass and fossil fuel burning.
5. Natural gas which is mainly methane. Unavoidable leakage from exploration and transportation add CH_4 to the atmosphere.

<u>Sinks</u>

Methane is lost from the atmosphere via the following sinks:

1. Reaction with OH• radical

$$\cdot OH + CH_4 \rightarrow \cdot CH_3 + H_2O \rightarrow \rightarrow CO_2 + H_2O$$

Recall that •OH is prevalent in the atmosphere and is basically produced by the reaction:

$$O^* + H_2O \rightarrow 2 \cdot OH$$

2. Absorption by soil
3. Loss to the stratosphere

The net result is that the steady state concentration of CH_4 hovers at about 1.7 ppm. The residence time of CH_4 in the atmosphere is about 15 years. On a molecule to molecule basis, CH_4 is about 20 times as effective a greenhouse gas as CO_2 because of the larger number of ways that the molecule can vibrate. However CO_2 is overall a greater contributor to global warming because of its much greater concentration.

Nitrous oxide, N_2O

Nitrous oxide (also called laughing gas because of its ability to initiate involuntary laughter) is another oxide of nitrogen in which the oxidation number of N is +1. (We have met NO and NO_2 before, in which the oxidation number of N is +2 and +4 respectively). The Lewis structure of N_2O is:

$$: N = N = O :$$

Global Warming

N_2O is a colorless gas. It is very stable and hardly reacts even at elevated temperatures. Its concentration has increased gradually since the 1600's and now stands at about 0.3 ppm in the atmosphere.

Sources

The following are likely sources of nitrous oxide:

1. Biological denitrification in anaerobic environments in which inorganic nitrate (NO_3^-) is mainly reduced to N_2 by certain bacteria but some N_2O is formed as a by-product.
2. Biological nitrification in aerobic environments in which ammonia (NH_3) and ammonium ion (NH_4^+) are oxidized first to the nitrite ion (NO_2^-), then to nitrate (NO_3^-), again with N_2O being a by-product

Sinks

There are no tropospheric sinks for N_2O because it is so chemically unreactive. Instead, it rises into the stratosphere where it is broken down by UV radiation.

Global Warming

To restate the phenomenon of global warming then, greenhouse gases at their 'normal' concentrations are necessary in order to maintain the earth's surface temperature at about 15°C by trapping IR (heat) radiation and re-radiating some of it back to earth. However, substantially higher concentrations of CO_2 and other greenhouse gases would trap more IR than usual, thereby returning more IR to earth (the enhanced greenhouse effect) and increasing earth's surface temperature.

The scientific evidence for global warming seems convincing. However, there are a number of people who are not convinced that increased concentrations of greenhouse gases have caused an increase in the temperature of the troposphere. Indeed, it is true that both the average global temperature and the concentration of greenhouse gases in the atmosphere have increased over the past one hundred years but the question is: Is there necessarily a cause-and-effect relationship? We have not heard the last word in this debate.

7 - Air Pollution

Effects of Global Warming

The major predicted effects from global warming are:

1. Less rainfall in temperate zones, for example, in the U.S. Midwest ('bread basket' regions) and the Canadian prairies. This will decrease agricultural output severely.
2. Increased rain fall in drier regions of Africa, India and the U.S. Southwest, thus increasing agriculture in these marginal areas depending on the quality of the soil.
3. Extension of agriculture to higher latitudes
4. Development of a tropical climate in the southern U.S.
5. Melting of polar ice caps resulting in a rise in sea level and inundation of coastal cities and extinction of some species.

Control Strategies for Global Warming

Some of the solutions which have been proposed to control global warming include:

1. Reduction of CO_2 emissions (the main culprit) by reducing fossil fuel use. An attempt at an international agreement was made in Kyoto, Japan in 1997 but many of the major players like the United States of America are balking because of the potential impact on their energy use. Even developing countries seem less than enthusiastic because they fear this would inhibit their economic development. European countries like England, France and Germany seem to take the issue of global warming more seriously and are at the forefront of the push for meaningful international agreement.
2. Switching to alternative fuels such as hydrogen (H_2) (in which the only by-product of combustion is H_2O) as well as other cleaner forms of energy such as solar and wind.

SELECTED REFERENCES

Sections 7.2 Photochemical Smog

1. J. M. Beard, " Chemistry, Energy and the Environment ", pp 111 - 121, Wuerz, 1995

2. C. Baird, " Environmental Chemistry ", pp 82 - 90, 125 - 133, Freeman, 1995

Section 7.3 Stratospheric Ozone Depletion

3. J. M. Beard, ibid., pp 146 - 151
4. C. Baird, ibid., pp 13 - 67

Section 7.4 Acid Rain

5. J. M. Beard, ibid., pp 121 - 132
6. C. Baird, ibid., pp 90 - 99

Section 7.5 Global Warming

7. J. M. Beard, ibid., pp 136 - 146
8. C. Baird, ibid., pp 149 - 187

ADDITIONAL REFERENCES

1. S. E. Manahan, " Environmental Chemistry ", Lewis Publishers, 2000
2. N. Bunce, " Environmental Chemistry ", Wuerz, 1994
3. D. B. Botkin, E. A. Keller, " Environmental Science ", Wiley, 1998
4. G. T. Miller, " Environmental Science ", Wadsworth, 1999

QUESTIONS

Section 7.1- 7.2 Polluted Air/Photochemical Smog

7.1 What is polluted air? Name three ways in which air can become polluted.

7.2 What is photochemical smog? What are the pre-conditions for photochemical smog formation? Write equations where possible.

7.3 Distinguish between primary and secondary pollutants. Give an example of each type of pollutant.

7.4 NO and CO are two primary pollutants found in automobile exhaust. How are they produced? Write equations where possible.

7.5 How does an automobile catalytic converter work? Write equations where possible.

7.6 Describe a strategy for reducing photochemical smog formation.

Section 7.3 - Stratospheric Ozone Depletion

7.7 Define the stratosphere. What is the approximate composition of the stratosphere?

7.8 How is ozone produced and destroyed in the stratosphere? Write equations where possible.

7.9 CFC's are implicated in the enhanced destruction of the ozone layer. What does CFC stand for? List three uses of CFC's before they were banned.

7.10 Choose any one CFC and write its formula. Show how it can be effective in the destruction of stratospheric ozone. Write equations for the mechanism.

7.11 Describe a strategy for reducing the destruction of the ozone layer.

Section 7.4 - Acid Rain

7.12 Define acid rain. Name two primary pollutants which can cause acid rain and briefly describe how they are produced.

7.13 What are the main sources of SO_2 in the atmosphere? Write equations to show the formation of SO_2 in the air.

7.14 How does acid rain affect the environment? Write equations where possible.

Questions

7.15 Describe a strategy for the control of acid rain.

Section 7.5 - Global Warming

7.16 What is infra-red radiation? What is its relationship to global warming?

7.17 What is a greenhouse gas? Name three greenhouse gases and write their formulas.

7.18 What is the main gas responsible for global warming and what are the main sources of this gas?

7.19 What are some of the predicted effects of global warming?

7.20 Describe a strategy for reducing global warming.

PART C - WATER POLLUTION

Chapter 8 - Water and Aqueous Solutions

8.1 Properties of Water

Water is a colorless, odorless and tasteless liquid at room temperature. It is a very stable substance and probably dissolves more substances than any other solvent. For this reason, it is often called the ' universal' solvent. Some of the properties that make it a unique liquid are :

(1) Intermolecular Forces in Water

We have seen how the atoms within a molecule are held together either by strong ionic or covalent bonds. But molecules themselves are attracted to other molecules. The forces between molecules are called intermolecular forces and are much weaker than intramolecular covalent and ionic bonds (20 kJ/mol vs 200 kJ/mol). Intermolecular forces in water (as in all liquids) are of three main types (a) dipole-dipole interactions (b) hydrogen bonding and (c) London dispersion forces.

(a) **Dipole-dipole** forces exist between polar molecules like water, in which the positive end of one polar molecule is attracted to the negative end of a second polar molecule. The forces are Coulombic in nature, much like those in ionic compounds, only weaker.

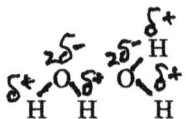

(b) **Hydrogen bonding** occurs between the H atom of a hydrogen-containing molecule like water and the O, N, F atom of the same molecule or a different molecule. Thus, besides dipole-dipole interactions, hydrogen bonding also occurs in pure water and together account for the unusually high boiling point of water. This is shown in Figure 8.1.

(c) **London dispersion** forces are the only intermolecular forces operating in nonpolar molecules. Even though nonpolar molecules do not have a permanent polarity, a given molecule may experience a distortion of its electron density distribution, for example, by radiation, and so

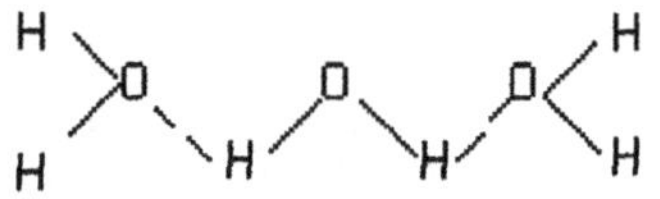

Intermolecular Forces

H–O(H) ⋯ H–O(H) ⋯ H–O(H) ⋯ H–O(H)

Figure 8.1 Hydrogen bonding in water

acquire an instantaneous dipole. This, in turn, induces a dipole in the adjacent nonpolar molecule, thus leading to an instantaneous but temporary attraction between the two molecules. London forces (sometimes called Van der Waals forces) are very weak and although they are the only forces in nonpolar liquids, they exist in all molecules.

(2) Vapor Pressure of Pure Water

Consider a pure solvent like H_2O, in a closed container held at a constant temperature and fitted with a pressure measuring device. The molecules of H_2O in the liquid phase are moving at random (though not as randomly as in a gas) and with kinetic energies such that there is a distribution of kinetic energies ranging from very low to very high values. Also the intermolecular forces of attraction are greater than in a gas. Those molecules which arrive at the liquid surface and have kinetic energies greater than the attractive (dipole-dipole and hydrogen bonding) energies, will escape from the liquid surface and become converted to vapor. This process is called vaporization and is a **phase transition**. The pressure measuring device will record an increase in gas pressure as more and more vapor molecules fill the space above the liquid.

Vapor molecules are also being returned to the liquid phase (a process called condensation) as more molecules leave the liquid. This dual

8 - Water and Aqueous Solutions

process will continue until the rates of vaporization and condensation become equal, **provided there is enough liquid water present**. The system is then at equilibrium and the pressure measuring device shows no further change in gas pressure. This equilibrium pressure (P°) exerted by the pure vapor at the stated temperature is called the vapor pressure of the pure liquid. For water at 298 K, P° = 23.8 torr.

The vapor pressure of a pure liquid is a function of temperature, increasing (exponentially) with an increase in (Kelvin) temperature. For water, P° = 760 torr at T = 373 K (100 °C). See Figure 8.2. The temperature of 373 K is therefore called the normal boiling point of water. Moreover, a liquid boils (evidenced by rising air bubbles) when its vapor pressure equals the external barometric pressure. Therefore, the boiling point of water depends on the existing barometric pressure.

Different liquids have different vapor pressure profiles which reflect their different intermolecular forces. For example, diethyl ether has a relatively high vapor pressure of 600 torr at 298 K and a relatively low normal boiling point of 308 K (35 °C) because its intermolecular forces are weak compared to those in water.

(c) Heat of Vaporization of Water

We have seen that if a quantity of heat q is added to a pure substance such as a liquid, very often it goes into raising the temperature of the liquid provided the temperature is below the boiling point of the liquid at the prevailing barometric pressure.

(The equation was: $q = mC \Delta T$

where C is the heat capacity of the substance and ΔT is the change in temperature which occurs). What happens if we continue to add heat at the boiling point of the liquid? Well, the temperature does not change but the heat goes directly into vaporizing the liquid at the boiling point. We then talk about the molar heat or enthalpy of vaporization (ΔH_v) of the liquid which is defined as the amount of heat (in J) necessary to vaporize one mole of the liquid at its boiling point

For the process:

$$H_2O\ (\ell) \rightarrow H_2O\ (g)$$

$$\Delta H_v = 2.26\ kJ/g\ or\ 40.7\ kJ/mol$$

Density of Water

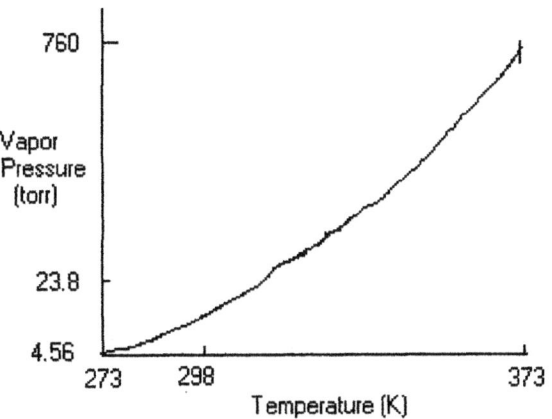

Figure 8.2 The temperature dependence of the vapor pressure of water

Similarly it takes an injection of heat to convert solid H_2O (ice) at its melting point of 0 °C to liquid H_2O at 0 °C. This is called the heat or enthalpy of fusion and has a value of 335 J/g or 6.02 kJ/mol.

(4) <u>Density of Water</u>

The density of a liquid (mass per unit volume) normally increases as the temperature decreases (since the volume decreases as the temperature decreases). The temperature dependence of the density of water is unique. Between 100 °C and 4 °C, water behaves normally in the sense that its density increases from 0.959 g/mL to 1.00 g/mL. However, below 4 °C, the density of water begins to decrease either as liquid or solid; water has its maximum density at 4 °C.

Ice is less dense (0.917 g/mL) at 0 °C than liquid H_2O at 0 °C because the H-bonding in ice orients the molecules in such a way that there are many open spaces, making it less dense. Therefore, ice floats on water and this fact has important consequences for the survival of fish and other aquatic species in lakes and oceans.

8 - Water and Aueous Solutions

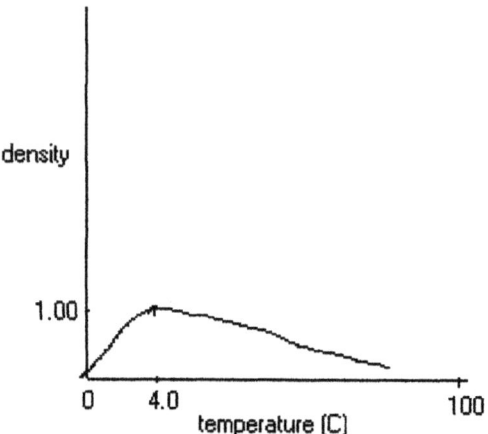

Figure 8.3   The variation of density of water with temperature

8.2   Aqueous Solutions

A solution may be defined as a homogeneous mixture of two (or more) substances with one of the substances (usually the one in larger amount) being the solvent and the other(s) the solute(s).

Classification of Solutions:
1.   Types of Phases Present

A number of types of solutions arise from the possible combinations of phases present -solid,liquid or gas. We may have solutions of gases in gases (air), gases in liquids (CO_2/H_2O or soda water, HCl_{aq}), gases in solids (H_2 in SiO_2), liquids in liquids (H_2SO_4/H_2O), solids in liquids ($NaCl/H_2O$), solids in solids (brass-Cu, 60%, Zn, 40%). For now, we will concentrate on solutions of solids in liquids , and in particular, solids called non-electrolytes whose solutions do not conduct electricity, for example, glucose/H_2O but not NaCl/ water.

2.   Relative Amounts of Solute and Solvent

Solutions may be classified depending on the relative proportion of

Concentration Terms

solute to solvent. A saturated solution contains the maximum amount of a given solute that will dissolve in a given quantity (say 100g) of solvent at a given temperature. Saturation is evidenced by the presence of solid in equilibrium with solution. We then talk about the solubility of the solute at a stated temperature. Units may g L^{-1} or mol L^{-1}. Note that solubility is, in general, affected by the temperature usually increasing, (e.g. sucrose/H_2O) but it may also decrease (Na_2SO_4/H_2O), gas/liquid (CO_2/H_2O), or may remain more or less the same over an extended temperature range (NaCl/H_2O). The exact behavior depends on whether the solution process is endothermic (increase), exothermic (decrease) or neither.

An unsaturated solution contains less than the maximum amount of solute that can be dissolved at the stated temperature. Notice we may go from the saturated to unsaturated by changing the temperature (depending on solubility/temperature curve) so that a given solution may be saturated at one temperature but unsaturated at a higher temperature.

A supersaturated solution at a given temperature has more than the maximum amount usually present in a saturated solution at that temperature. It is a very unstable (metastable) system; many solutions on cooling remain supersaturated for a while but precipitates solid on being agitated.

8.3 Concentrations of Solutions

Recall that a solution is a homogeneous mixture of two or more substances. For a binary solution (that is, one made up of two components), the substance in larger amount is usually called the solvent and the other is the solute. A number of different but related quantities are used to describe the concentration of a solution which tells us how much solute and solvent are present in the solution.

(a) Mass Fraction and Mole Fraction

Mass (weight) fraction of a component is the ratio of the mass (in grams) of that component to the total mass (in grams) of all substances present. Thus for a binary solution containing components 1 and 2

$$x_1 = \frac{w_1}{w_1 + w_2}, \quad x_2 = \frac{w_2}{w_1 + w_2} \quad \text{and} \quad x_1 + x_2 = 1$$

8 - Water and Aqueous Solutions

Multiplying by 100 converts the mass fraction into percent by mass (weight) which is sometimes written as % w/w. A related measurement of concentration especially for very dilute solutions is ppm or parts per million. 1 ppm represents 1g of the solute in 1 million g of the solution or 1 µg of solute in 1mL of solution or 1 mg in 1 L.

(b) <u>Mole Fraction</u>

Similarly the mole fraction is defined with respect to the mole

$$X_1 = \frac{n_1}{n_1 + n_2} \quad , \quad X_2 = \frac{n_2}{n_1 + n_2} \quad , \quad X_1 + X_2 = 1$$

where n_1 and n_2 are the number of moles of components 1 and 2. Multiplying by 100 converts mole fraction into mol %.

(c) <u>Molarity</u>

The molarity M_i of a solute in a solution is the number of moles of solute per liter of solution.

$$M_1 = \frac{n_1}{V}$$

where V is the total volume of the solution.

<u>Example</u>

A vinegar solution (an aqueous solution of acetic acid, $C_2H_4O_2$) contains 5.0 g acetic acid in 95 g water. What is the weight percent of acetic acid in vinegar and the molarity of the solution? Assume the density of the solution is 1.0 g/mL.

Let us label acetic acid 1 and water 2. Then

$$m_1 + m_2 = 5.0 + 95 = 100 \text{ g}$$

Weight % of aectic acid = $\frac{5.0 \text{ g}}{100 \text{ g}} \times 100 = 5.0$

Also 5.0 g × $\frac{1 \text{ mol acetic}}{60 \text{ g}}$ = 0.083 mol acetic acid

$$V = \frac{m}{d} = \frac{(100 \text{ g}) \text{ mL}}{1.0 \text{ g}} = 100 \text{ mL} \times \frac{1 \text{ L}}{10^3 \text{ mL}} = 0.10 \text{ L}$$

$$M = \frac{n_1}{V} = \frac{0.083 \text{ mol}}{0.10 \text{ L}} = 0.83 \text{ mol/L}$$

8.4 Colligative Properties of Solutions

Let us now consider the properties of an aqueous solution containing a solute that does not have a measurable vapor pressure at the stated temperature. The solute is called a non-volatile solute. The vapor pressure of the solution is then due only to solvent molecules. The properties of such an aqueous solution depend only on the concentration of the added non-volatile solute and not on the nature of the solute. These properties are therefore called colligative (collective) properties of solutions. Some of these properties are:

(a) **Vapor Pressure Lowering**

The vapor pressure of a solution containing a non-volatile solute is lower than the vapor pressure of the pure solvent at the same temperature. The vapor pressure of the solution depends on the concentration of the solution and scales with the mole fraction of the solute; that is, the higher the mole fraction of the solute, the lower the vapor pressure of the solution.

(b) **Boiling Point Elevation**

Consider again water in a closed container fitted with a pressure release valve that pops open when the vapor pressure of the liquid is equal to the external barometric pressure P of 1 atm. Let us now increase the temperature slowly from room temperature and a barometric pressure of 1 atm and examine what happens to the vapor pressure If the water is pure, the vapor pressure continues to rise until at 100 °C, the water boils and the valve pops open indicating that the vapor pressure is equal to the barometric pressure of 1 atm.

Now, let us replace the pure water with a 0.10 M solution of glucose in water and repeat the experiment. As we have seen before, the non-

8 - Water and Aqueous Solutions

volatile glucose will lower the vapor pressure of the water at any given temperature so that at 100 °C, we find that the pressure of the vapor is **less** than 1 atm and the valve does not pop open. We need to increase the temperature to about 100.2 °C for the vapor pressure to reach 1 atm and therefore, cause the water to boil and the valve to pop open In other words, the boiling point of the aqueous solution is **raised** relative to that of pure water, by the addition of the non-volatile solute.

The amount by which the boiling point is raised (called the boiling point elevation, ΔT_b) also depends on the concentration of the solution and increases as the mole fraction of the solute increases. Engine coolant is basically a mixture of water and a non-volatile liquid (ethylene gycol) and its boiling point is therefore higher than the boiling point of pure water.

(c) Freezing Point Depression of Water

The addition of a non-volatile solute to water also causes a lowering or depression of the freezing point of the solution relative to that of pure water. Again, the magnitude of the depression (ΔT_f) depends on the concentration of the solution; the higher the mole fraction of the solute, the lower the freezing point of the solution. One practical aspect of this effect occurs when we add salt to an icy road in order to cause the ice to melt because its melting/freezing point has been lowered.

8.5 Acids, Bases and Salts

Arrhenius Concept of Acids/Bases

According to Arrhenius (1884), an acid is a hydrogen-containing compound that dissolves in water to produce hydrogen ions (H^+). For example, HNO_3 is an Arrhenius acid because of the following reaction in water:

$$HNO_3 \text{ (aq)} \rightarrow H^+ \text{ (aq)} + NO_3^- \text{ (aq)}$$

On the other hand, an Arrhenius base is a substance which dissolves in water to produce OH^- (hydroxide) ions, for example, NaOH:

$$NaOH(s) \xrightarrow{H_2O} Na^+(aq) + OH^-(aq)$$

Acids and Bases

The reaction between an Arrhenius acid and a base produces a salt and water and is called neutralization.

$$NaOH\ (aq) + HNO_3\ (aq) \rightarrow NaNO_3\ (aq) + H_2O\ (aq)$$

Note that the only species that have changed are H^+ and OH^- since Na^+ and NO_3^- were originally present in the aqueous solutions of the reactants. Neutralization may therefore be written as

$$H^+\ (aq) + OH^-\ (aq) \rightarrow H_2O\ (l)$$

Bronsted - Lowry Concept of Acids/Bases

The Bronsted-Lowry (1923) concept of acids and bases regards an acid as a proton donor and a base as a proton acceptor. Thus HF is a Bronsted acid and H_2O can function as a Bronsted base as the following equation shows:

$$HF(g) + H_2O(l) \rightarrow H_3O^+(aq) + F^-(aq)$$
(acid)   (base)

The fact is that the reverse reaction also occurs:

$$H_3O^+ + F^- \rightarrow HF + H_2O$$
(acid)   (base)

and H_3O^+ (the hydronium ion) is a Bronsted acid and F^- is a Bronsted base.

Note then that we have, in this acid-base reaction, two pairs, (HF/F^-, and H_3O^+/H_2O), the members of each pair differing by only a proton. Each pair is called a conjugate pair, one member of which is the conjugate acid and the other member the conjugate base.

$$HF\ +\ H_2O \rightleftharpoons H_3O^+ + F^-$$
acid 1 + base 2   acid 2 + base 1

One more example of a Bronsted acid-base reaction is

$$NH_3(g) + H_2O(l) \rightarrow NH_4^+\ (aq) + OH^-\ (aq)$$

8 - Water and Aqueous Solutions

Notice in this case NH_3 is a Bronsted base and its corresponding Bronsted acid is NH_4^+. H_2O is acting as a proton donor or Bronsted acid and OH^- is its conjugate base. Thus water is said to be amphiprotic because it can act both as an acid and a base. A consequence of this is that H_2O itself ionizes to give small amounts of hydronium and hydroxide ions.

$$H_2O\,(l) + H_2O(l) \rightleftharpoons H_3O^+\,(aq) + OH^-(aq)$$

or

$$H_2O \rightleftharpoons H^+ + OH^-$$

Notice that the Bronsted-Lowry definition also includes Arrhenius acids and bases.

Properties of Acids

We can summarize the behavior of acids as follows:

1. They dissolve in H_2O to produce H^+ ions
2. They have a sour taste
3. They react with an oxide or hydroxide to produce a salt and water

$$CuO(s) + 2\,H^+(aq) \rightarrow Cu^{2+}(aq) + H_2O(l)$$

$$Mg(OH)_2(s) + 2\,H^+(aq) \rightarrow Mg^{2+}(aq) + 2\,H_2O(l)$$

4. They react with a carbonate or bicarbonate to produce CO_2, a salt and water

$$CaCO_3(s) + 2\,H^+(aq) \rightarrow Ca^{2+}(aq) + CO_2(g) + H_2O(\ell)$$

$$HCO_3^-(aq) + H^+(aq) \rightarrow CO_2(g) + H_2O(\ell)$$

5. They react with metals that are stronger reducing agents than H_2 and in the process, H_2 is displaced from the acid

$$Zn(s) + 2\,H^+ \rightarrow Zn^{2+}(aq) + H_2(g)$$

Properties of Bases:

1. They dissolve in H_2O to produce OH^-

Salts

$$Na_2O(s) + H_2O(l) \rightarrow 2\ Na^+\ (aq) + 2\ OH^-(aq)$$

2. They have a bitter taste and a soapy feel

3. They react with acids to produce a salt and water

$$Na_2O(s) + 2\ H^+\ (aq) \rightarrow 2\ Na^+\ (aq) + H_2O(l)$$

Salts

A salt is a substance which consists of a positive ion or **cation**, such as a metal ion or NH_4^+, and a negative ion or **anion** which may be monatomic (Cl^-) or polyatomic (CO_3^{2-}).

Salts are formed by reaction of an acid with a metal, metal hydroxide, metal oxide, metal carbonate or metal bicarbonate. For example, the salt magnesium chloride ($MgCl_2$) may be prepared by any of the following methods:

1. $Mg(s) + 2\ HCl(aq) \rightarrow MgCl_2(aq) + H_2(g)$

2. $Mg(OH)_2(s) + 2\ H^+(aq) \rightarrow Mg^{2+}\ (aq) + 2\ H_2O(l)$

3. $MgO(s) + 2\ H^+(aq) \rightarrow Mg^{2+}\ (aq) + H_2O(l)$

4. $MgCO_3(s) + 2\ H^+(aq) \rightarrow Mg^{2+}\ (aq) + CO_2(g) + H_2O(l)$

5. $Mg(HCO_3)_2 + 2\ H^+ \rightarrow Mg^{2+} + 2\ CO_2(g) + 2\ H_2O(l)$

8.6 Strength of Acids and Bases

Let us look at the reaction of HCl gas with H_2O according to the Bronsted-Lowry concept:

$$HCl(g) + H_2O(l) \rightarrow H_3O^+(aq) + Cl^-(aq)$$

or more simply as:

$$HCl(aq) \rightarrow H^+\ (aq) + Cl^-\ (aq)$$

8 - Water and Aqueous Solutions

What we find is that HCl reacts more or less completely with H_2O to form H^+ ions and Cl^- ions and that there are practically no neutral, discrete HCl molecules in the aqueous solution, only ions. We therefore call HCl a strong Bronsted-Lowry acid since its aqueous solution consists essentially of ions.

On the contrary, consider the reaction of acetic acid with H_2O:

$$CH_3COOH(aq) \rightleftharpoons H^+(aq) + CH_3COO^-\ (aq)$$
(acetic acid) \hspace{2cm} (acetate ion)

Here, only a small fraction (less than 1 %) of the neutral CH_3COOH molecules are converted to ions; there is still a considerable number of discrete molecules of CH_3COOH remaining in solution at equilibrium. Acetic acid is therefore termed a weak acid. The majority of Bronsted-Lowry acids are weak, with the six strongest acids being HCl < H Br < HI < HNO_3 < H_2SO_4 < $HClO_4$. Strangely enough, HF is a weak acid.

Similarly, strong bases dissolve in water generating large numbers of OH^- ions. (Recall that Bronsted bases accept protons). NaOH is already made up of Na^+ and OH^- ions in the solid state and dissolving in water simply pulls the ions apart so that NaOH is practically completely ionized in water.

$$NaOH\ (aq) \rightleftharpoons Na^+\ (aq) + OH^-\ (aq)$$

On the other hand, Na_2O which contains Na^+ and O^{2-} ions in the solid state, reacts with water to form OH^- ions (O^{2-} is a proton acceptor and hence a Bronsted base)

$$Na_2O\ (s) + H_2O(l) \rightleftharpoons 2\ Na^+\ (aq) + 2\ OH^-(aq)$$

Or \hspace{2cm} $O^{2-} + H_2O \rightleftharpoons 2\ OH^-$

NaOH and Na_2O are both considered strong bases, as are NaH and $NaNH_2$ containing the H^- and NH_2^- anions.

Similarly weak bases react with water by removing protons from water only to a very small extent; for example, NH_3 is a weak base because its aqueous solution contains relatively few OH^- ions.

$$NH_3(g) + H_2O\ (\ell) \rightleftharpoons NH_4^+(aq) + OH^-(aq)$$

8.7 The pH Scale

We have said before that H_2O self-ionizes as follows:

$$2 H_2O(\ell) \rightleftharpoons H_3O^+(aq) + OH^-(aq)$$

or $\quad H_2O(\ell) \rightleftharpoons H^+(aq) + OH^-(aq)$

but to what extent? That is, what is the concentration of H^+ and OH^- in a pure water sample? Careful measurements, such as electrical conductivity, have shown that the concentration of H^+ (C_{H+}) is 1.0×10^{-7} mol L^{-1} at 25 °C. This is also the concentration of OH^-, from the stoichiometry of the ionization reaction. We may write then:

$$K_w = [H^+][OH^-] = (1.0 \times 10^{-7})(1.0 \times 10^{-7}) = 1.0 \times 10^{-14}$$

where $[H^+]$ is the concentration of H^+ relative to some standard concentration, $C°$ which is taken as 1 mol L^{-1}, thus making $[H^+]$ <u>unitless</u> but numerically equal to C_{H+}.

K_w is called the ionic product of water. In an aqueous solution of any substance, this relationship between H^+ and OH^- must be satisfied regardless of the relative amounts of H^+ and OH^- in the solution.

Example:

What is the unitless concentration of H^+ [H^+] and OH^- [OH^-] in an aqueous solution of HCl in which 0.01 mol of HCl is dissolved in 1 L of solution?

$$HCl(aq) \rightarrow H^+(aq) + Cl^-(aq)$$

Basis: 1 L
Initial   0.01 mol   0   0
Final   0   0.01 mol   0.01 mol

$$[H^+] = \frac{C_{H+}}{C°} = \frac{1.0 \times 10^{-2} \text{ mol } L^{-1}}{1 \text{ mol } L^{-1}} = 1.0 \times 10^{-2}$$

$$[OH^-] = \frac{K_w}{[H^+]} = \frac{1.0 \times 10^{-14}}{1.0 \times 10^{-2}} = 1.0 \times 10^{-12}$$

8 - Water and Aqueous Solutions

Also $\quad C_{OH^-} = C° [OH^-] = (1 \text{ mol L}^{-1})(1.0 \times 10^{-12}) = 1.0 \times 10^{-12} \text{ mol L}^{-1}$

Rather than dealing with negative powers of 10 to represent $[H^+]$ and $[OH^-]$, chemists have adopted more convenient quantities which are normally positive and lie between 0 and 14. They are called the pH and pOH and are defined as follows:

$$pH = -\log_{10}[H^+] \quad \text{or} \quad [H^+] = 10^{-pH}$$

and

$$pOH = -\log_{10}[OH^-] \quad \text{or} \quad [OH^-] = 10^{-pOH}$$

Example:

Calculate the pH and pOH of the above-mentioned solution of HCl in water.

$[H^+] = 1.0 \times 10^{-2}$
$\log[H^+] = \log(1.0) + \log(10^{-2}) \quad (\log ab = \log a + \log b)$
$\quad\quad\quad\quad = 0 + (-2) = -2$

$pH = -\log[H^+] = -(-2) = 2$

Alternatively, by comparison, $[H^+] = 10^{-pH} = 10^{-2}$ or pH = 2

Similarly: $\quad [OH^-] = 1.0 \times 10^{-12}$
$\quad\quad\quad\quad \log[OH^-] = 0 + (-12) = -12$
$\quad\quad\quad\quad pOH = -\log[OH^-] = -(-12) = 12$

Note here that:

$$pH + pOH = 2 + 12 = 14.$$

This is also generally true since:

$[H^+][OH^-] = 1.0 \times 10^{-14}$
$\log[H^+] + \log[OH^-] = -14$
$-\log[H^+] + (-\log[OH^-]) = 14$
or $\quad pH + pOH = 14$

Measurement of pH

Normally pH (and pOH) lies between 0 and 14 although pH could be less than 0 or greater than 14.

As we have seen before, pure water contains $[H^+] = [OH^-] = 1.0 \times 10^{-7}$ so that pH = pOH = 7. Hence, pure water is said to be neutral. Solutions whose pH < 7 (pOH > 7) are said to be acidic whereas solutions whose pH > 7 (pOH < 7) are said to be basic. Note, however, that both types of solutions contain H^+ and OH^- ions; they differ only in the relative proportions of H^+ and OH^-.

Measurement of pH

The pH of solutions is measured by two methods: (a) indicators/pH paper, (b) pH meters.

(a) Indicators

Some organic compounds called dyes have definite colors depending on the acidity or basicity of their aqueous solutions. Different dyes have different colors over different pH ranges; for example, litmus indicator changes color from red to blue over a pH range of 4.7- 8.3. In the universal indicator paper for measuring pH, pieces of paper are impregnated with a mixture of indicators. A color chart shows the color corresponding to a certain pH range, e.g. 1-2, 2-3, etc.

To measure the pH of a given solution, a piece of paper is treated with a drop or two of the solution and the paper changes color. Its color is compared to that of the color chart, thereby establishing an approximate pH for the solution.

(b) pH Meter

Accurate measurements of pH are done using a pH meter. This has a glass electrode whose voltage depends on the pH of the solution. A meter gives a direct read out of the pH from the voltage of the glass electrode. Before use, the pH meter must be calibrated, that is, the electrode is dipped into at least two solutions of known pH.

Needless to say, the acidity or basicity of many systems plays a very important role in the operation of these systems, e.g. many biochemical reactions in the body take place at nearly neutral pH (7) and delicate processes occur to maintain the pH at this value. Figure 8.4 shows the pH of some commom substances. For example, human serum (blood

8 - Water and Aqueous Solutions

plasma) has a pH of 7.3 - 7.6 and is therefore slightly basic. Human saliva varies from slightly acidic to slightly basic (pH of 6.5 - 7.5). Car battery acid has a pH of about 1, stomach acid about 1.0- 3.0, vinegar about 2.5, rain water about 5.6, sea water about 7.5 - 8.5, baking soda about 9 and milk of magnesia about 10.

8.8 Buffers

We have mentioned that the pH of the blood is maintained within a fairly narrow range of about 7.3 - 7.5 This is due to the presence in blood of a mixture of substances called a buffer system, specifically the HCO_3^-/CO_3^{2-} system. A buffer is defined as a mixture which will resist a

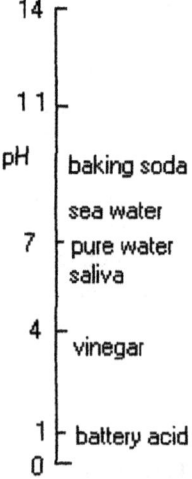

Figure 8.4 The pH of some common substances

change of pH when small amounts of acid or base are added to it.

Nature of Buffers

Buffers are usually either (a) a solution of a weak acid and a soluble salt of the weak acid (for example, a mixture of CH_3COOH and $Na^+CH_3COO^-$) and (b) a solution of a weak base and a soluble salt of the weak base (for example, NH_3/NH_4Cl mixture). For the first buffer system, the following equations may be written if we represent CH_3COO^- by A^-:

$$HA(aq) \rightleftharpoons H^+ + A^-$$

$$Na^+A^-(aq) \rightarrow Na^+(aq) + A^-(aq)$$

Notice that (i) A^- is common to both species - the common ion effect (ii) the soluble salt, Na^+A^-, is completely ionized in aqueous solution while HA is only slightly ionized and so the concentration of A^- in the mixture is essentially that of the Na^+A^- salt.

How does this buffer work? The addition of a fairly small quantity of H^+ ions to the buffer results in its immediate removal by combining with A^- ions to form more neutral HA molecules by the backward reaction, according to LeChatelier's principle.

$$CH_3COO^- + H^+ \rightarrow CH_3COOH$$

Addition of base results in its removal by reaction with HA

$$CH_3COOH + OH^- \rightarrow CH_3COO^- + H_2O$$

Hence, the original acidity/basicity is more or less maintained within a fairly narrow range when small amounts of acid or base are added to it. The pH of a particular buffer system at a given temperature depends only on the **ratio** of the salt concentration to the acid or base concentration.

Capacity of Buffers

When dealing with buffers, we are often concerned not only with its pH as discussed above but also with its **buffering capacity**. The buffering capacity is the amount of acid or base the buffer is able to neutralize before the pH changes appreciably, and this depends on the **absolute** amounts of acid and base present in the buffer. Thus a solution that is 1.0 M in each of CH_3COOH and $Na^+CH_3COO^-$ will have the same pH as a solution that is 0.10 M in each of CH_3COOH and $Na^+CH_3COO^-$ since the ratio of salt to acid concentration is the same in both cases.

8 - Water and Aqueous Solutions

However, the former solution will have a greater buffering capacity because of its greater concentration.

SELECTED REFERENCES

Section 8.1 - Properties of Water

1. T. L. Brown, H. E. LeMay, B. E. Bursten, " Chemistry - The Central Science ", pp 393 - 411, Prentice-Hall, 2000
2. M. S. Silberberg, " Chemistry", pp 422 - 447, McGraw-Hill, 2000

Sections 8.2 - 8.4, Properties of Solutions

3. Brown et al., ibid. pp 469 - 495
4. M. S. Silberberg, ibid., pp 480 - 506

Sections 8.5 - 8.8 ,Acid, Base and Salts, pH, Buffers

5. Brown et al., ibid. pp 593 - 618, 641 - 650
6. M. S. Silberberg, ibid., pp 756-772, 805-816

QUESTIONS

Section 8.1 - Properties of Water

8.1 Describe two types of intermolecular forces operating in pure liquids. Give two examples of molecules containing these forces.

8.2 List and describe the intermolecular forces present in the following molecules (a) NH_3 (b) CH_4 (c) HBr

8.3 What is meant by the vapor pressure of a liquid? How is it qualitatively related to the boiling point of the liquid?

8.4 Define the normal boiling point of a liquid. Explain why water has an unusually high normal boiling point.

8.5 What is meant by the heat of vaporization of a liquid? Explain why water has an unusually high heat of vaporization.

Questions

8.6 Define the density of a liquid and describe how the density of a liquid normally varies with temperature. How is water different from most liquids in its density behavior.

Section 8.2 - Aqueous Solutions

8.7 Define the terms solution, saturated solution, solubility.

8.8 Distinguish between a saturated solution and a supersaturated solution. How can a saturated solution be converted into a supersaturated solution and vice versa?

Section 8.3 - Concentrations of Solutions

8.9 Name and define three ways of measuring the concentration of a solution.

8.10 Define the molarity of a solution. What is the molarity of a solution of ascorbic acid (vitamin C, $C_6H_8O_6$) which contains 1.76 g of the acid in 100 mL of water?

Section 8.4 - Colligative Properties of Solutions

8.11 What is a colligative property as applied to a solution? Name two colligative properties.

8.12 What effect does increasing the concentration of a non-volatile solute in water have on the boiling point and freezing point of water?

Section 8.5 - Acids, Bases and Salts

8.13 What is an acid according to the Arrhenius concept? Give an example. List three properties of acids. Write equations where possible.

8.14 What is a base according to the Bronsted-Lowry concept? Give an example. List three properties of bases. Write equations where possible.

8 - Water and Aqueous Solutions

8.15 Define a salt. Give three methods for the preparation of salts. Write equations for the reactions.

Section 8.6 - Strength of Acids and Bases

8.16 Distinguish between a strong acid and a weak acid. Give an example in each case.

8.17 Distinguish between a strong base and a weak base. Give an example in each case.

8.18 Label each of the following as a strong acid/base or weak acid/base (a) NH_3 (b) HCl (c) KOH (d) CH_3COOH. Say why in each case.

Section 8.7 - The pH Scale

8.19 Write an equation to show the self-ionization of water. Write an expression for the ionic product of water, K_w.

8.20 What is the $[H^+]$ of a solution whose $[OH^-]$ is 10^{-8}? Is the solution acidic or basic?

8.21 Define pH and pOH. What is the pH and pOH of a solution whose $[H^+]$ is 10^{-3}? Is the solution acidic or basic?

Section 8.8 - Buffers

8.22 Define a buffer and give an example.

8.23 A particular buffer is prepared by mixing a solution of acetic acid (CH_3COOH) and a solution of sodium acetate ($Na^+CH_3COO^-$). What happens when small amounts of either HCl or NaOH solution are added to the buffer? Write equations where possible.

8.24 What is meant by the buffering capacity of a buffer? What determines buffer capacity? What determines the pH of a buffer?

Chapter 9 - Water Pollution

9.1 Disribution of Water

Water is the most abundant compound on earth. We mentioned earlier the water (hydrologic) cycle in which water leaves the oceans, lakes/rivers and the land and enters the atmosphere. It is then returned to the earth as rain or snow. Water covers about two-thirds of the earth's surface. However, the distribution of water over the surface of the earth is far from uniform. The oceans are the main repository (> 97%) followed by lakes and ponds, rivers and streams, the atmosphere, and finally ground water. Other sources are glaciers and soil moisture.

Oceans

Even though about 97% of water is in the oceans, it is not generally useful for industry/ agriculture/ domestic use because of relatively high concentrations of dissolved solids. The composition of sea water is roughly as follows:

Table 9.1 Composition of Ocean Water

Ions	C (mmol L-1)	ppm
Cl^-	550	19,000
Na^+	460	10,600
Mg^{2+}	54	2,600
SO_4^{2-}	28	1,300
K^+	10	380
Ca^{2+}	10	400
HCO_3^-	2	140

As you can see, sea water is mostly NaCl at a fairly high concentration of about 0.5 M. One of the consequences of this is that sea water freezes at temperatures well below 0 °C. (Recall that the freezing point of a solvent like water is depressed by the presence of a non-volatile solute like NaCl). This is very important for the survival of fish and other marine life during the winter.

One other interesting fact about ocean water is that it has a long residence time which is about 3,000 years for surface water and 30,000 years for deeper water. This means that if the ocean becomes polluted, it will take a very long time to rid itself of the pollutants. Potable (drinking)

9 - Water Pollution

water may be obtained from sea water by a process called desalination.

Lakes

Lakes are an important source of relatively fresh water even though some lakes can be very salty. The increased salinity is due to evaporation and in fact, naturally occurring deposits of NaCl (rock salt) were formed through evaporation of ancient seas and lakes. The Great Lakes of Canada and the U.S. contain about 20% of the world's fresh water.

Rivers

Even though they represent only a small percent of surface water, rivers have historically been very important. Whole civilizations have grown up along rivers, for example, Egypt along the Nile River. Rivers like the Nile have been used for transportation, domestic drinking water and irrigation, among other things.

The composition of river water is as follows:

Table 9.2 Composition of River Water

Ion	C (in mmol L^{-1})
Cl^-	0.22
Na^+	0.27
Mg^{2+}	0.17
SO_4^{2-}	0.12
K^+	0.06
Ca^{2+}	0.04
HCO_3^-	0.10

Note that rivers contain almost the same dissolved solids as sea water in approximately the same ratio but at much lower absolute concentrations.

The residence time of water in rivers is relatively short (about 30 days) because the water is fast moving. As a result, a river can quickly cleanse itself of pollutants provided there is not a continuous resupply of pollutants.

Irrigation utilizes a lot of water from rivers and streams. One problem though is the increase in salinity of the soil as the water evaporates repeatedly, making the land eventually uncultivable. The river itself may become increasingly saline and desalination becomes necessary in order to reuse the river's water.

Ground water

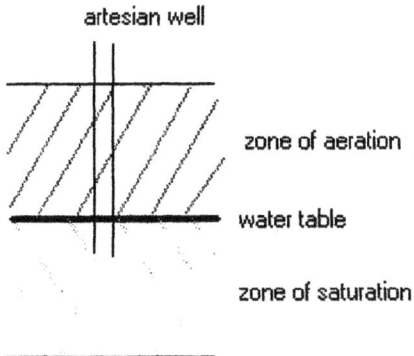

Figure 9.1 The composition of an aquifer

This originates from water reaching the earth as rain. The water percolates through the soil until it encounters an impermeable rock layer. Immediately above the rock layer is a zone of soil and water called the zone of saturation. The top of that layer is the water table. Immediately above the water table is a zone made up of soil, air and some water called the zone of aeration. The whole assembly is called an aquifer as illustrated in Figure 9.1

Aquifers are tapped with wells (artesian) which run straight down to the zone of saturation and serve as important sources of fresh water. Aquifers may be unconfined or confined depending on the area of resupply. If confined, the resupply area is small and the water is quickly depleted.

Purification of Drinking Water

Water from any source must often be treated before it can be used for

9 - Water Pollution

domestic purposes. The main steps in the purification of water are as follows:

1. **Aeration** removes dissolved gases like H_2S and volatile organic compounds (VOC's). It also causes oxidation of organic matter to CO_2 and water, and soluble iron (II) salts to insoluble iron (III) which precipitates as iron (III) hydroxide, $Fe(OH)_3$ and can be filtered off.

2. **Softening** of hard water by the addition of carbonate as Na_2CO_3 which precipitates Ca^{2+} and Mg^{2+} as their carbonates and hydroxides.

3. **Coagulation** of colloidal particles by addition of iron (III) sulfate and aluminum sulfate which form insoluble $Fe(OH)_3$ and $Al(OH)_3$ respectively thereby facilitating coagulation.

4. **Disinfection** by O_3, ClO_2 or HOCl to remove bacteria and other micro organisms. HOCl is obtained either from chlorinated water

$$Cl_2 + H_2O \rightarrow HOCl + HCl$$

or added as hypochlorite [$Ca(OCl)_2$ or NaClO]

$$OCl^- + H_2O \rightarrow HOCl + OH^-$$

The pH is maintained slightly above 7. Disinfection by Cl_2 sometimes forms organochlorines in the presence of organic matter. Organochlorines such as chloroform $(CHCl_3)$ are toxic. The limit of $CHCl_3$ in drinking water is 100 ppb. Disinfection is sometimes accomplished by UV radiation

5. **Membrane filtration** is especially useful in removing dissolved organic substances. The water is pumped through semipermeable membranes that allow passage of water but restrict organic molecules (a process called reverse osmosis).

9.2 Polluted Water

We are aware that pollutants in water can have catatrophic effects on the well- being of society. For example, micro-organisms in fecal matter that get into a domestic water supply may cause human diseases like cholera, dysentry, tubercolosis. But how do we define polluted water ?

Water Pollutants

The concept is not easily defined in absolute terms because, as we have seen, all natural water, whether in the oceans, rivers, lakes or in aquifers, contain certain impurities. Therefore, whether water is polluted or not depends on its end use. For example, the requirements for 'purity' of drinking water are not the same as those for irrigation water. Therefore, polluted water may be defined as water which is made unfit for a desired purpose.

Pollutants enter waterways from **point sources** such as sewage treatment plants, or **non-point sources** such as farm run-off. We will discuss water pollution in terms of the types and sources of pollutants, their chemistry and control strategies. Also, the treatment of waste water will be discussed.

The most common water pollutants are:

1. **Plant nutrients** such as phosphates and nitrates
2. **Heavy metals** such as lead (Pb), mercury (Hg) and cadmim (Cd)
3. **Organic chemicals** including dioxin, PCBs, pesticides and organic solvents such as trichloroethylene.
4. **Organic waste** from sources such as paper mills and food processors
5. **Other**, for example, disease-causing microorganisms (pathogens).
   This is more biological than chemical so we will gloss over this topic. They arise from human/animal waste entering water supplies. Diseases include cholera, typhoid fever, dysentery, hepatitis, polio.

9.3 Phosphates and Nitrates in Water

Phosphates

Phosphates are salts derived from the acid, orthophosphoric acid (H_3PO_4) which contains three ionizable hydrogens when dissolved in water and therefore form three types of salts: dihydrogen phosphate ($H_2PO_4^-$), hydrogen phosphate (HPO_4^{2-}) and orthophosphate (PO_4^{3-}).

$$H_3PO_4 (aq) \rightarrow H^+ + H_2PO_4^-$$
$$H_2PO_4^- (aq) \rightarrow H^+ + HPO_4^{2-}$$
$$HPO_4^{2-} (aq) \rightarrow H^+ + PO_4^{3-}$$

We shall represent all phosphates in the discussion that follows, by PO_4^{3-}

9 - Water Pollution

Sources

(a) As you know, plants manufacture their own food by photosynthesis using CO_2 from the air and H_2O from the soil. However they also need nitrogen (N), sulfur (S) and phosphorus (P) which are normally present as nitrates (NO_3^-), sulfates (SO_4^{2-}) and phosphates (PO_4^{3-}). These nutrients are often supplied as NPK fertilizers containing, for example, $(NH_4)_2SO_4$ which provides N and S, $Ca_3(PO_4)_2$ which provides P, and KNO_3 which provides K. Of course, the fertilizer is not all used up by the plants because farmers usually apply an excess of fertilizer. Some of the excess finds its way into rivers and lakes as run-off from farms, lawns, flower gardens and other activities which use fertilizers. These are all non-point sources.

(b) Phosphate also enters the water supply system by way of synthetic detergents (though this source has been significantly reduced). Detergents (as well as soaps) contain a long carbon chain that is non-polar forming the 'tail'; and a polar head such as $-SO_3^-Na^+$. The detergent acts by using the non-polar tail to bind to the oil or grease which is itself non-polar (a process called hydrophobic bonding) thus causing 'wetting' or dispersal of oil in the water during laundering. The polar end of the detergent is oriented towards the polar water molecules.

Tripolyphosphates (TP) were once added to detergents as 'builders' to tie up any Ca^{2+} and Mg^{2+} ions in the 'hard' water as water-soluble complexes thus allowing the detergent to do its work. They have been largely replaced by other 'builders' such as zeolites. The polyphosphate also made the water alkaline, which helped to keep dirt suspended in the water.

The tripolyphosphate ion is $P_3O_{10}^{5-}$ (the ion present in adenosine triphosphate (ATP) which is the major storehouse of chemical energy in living cells). The reaction with Ca^{2+} may be represented as:

$$P_3O_{10}^{5-}(aq) + Ca^{2+}(aq) \rightarrow [CaP_3O_{10}]^{3-}(aq)$$

When wash water containing sodium tripolyphosphate (STP) enters waterways, the excess STP is hydrolyzed to produce PO_4^{3-}

$$P_3O_{10}^{5-} + 2 H_2O \rightarrow 3 PO_4^{3-} + 4 H^+$$

(c) Another source of phosphates is untreated and treated human waste as well as animal waste.

Phosphates in Water

(d) Phosphate mining

What happens when PO_4^{3-} enters the waterways, for example, Lake Erie in 1960's? PO_4^{3-} is usually the 'limiting' (or controlling) nutrient for algal growth in lakes. Therefore, if the supply of PO_4^{3-} is increased above normal, algal growth increases exponentially and may cover large surface areas of the lake (a process called eutrophication). When the excess algae, (usually that below the surface) die, they decompose by oxidation, the water becomes depleted of dissolved oxygen (DO) and fish and other aquatic life are adversely affected. The lake water becomes foul-tasting and foul-smelling, green and slimy (with lots of dead fish).

Strategy for Control of Phosphate Contamination

A control strategy for PO_4^{3-} contamination is two-fold :

(1) Elimination at the source, for example, in detergents by using other builders, such as zeolites or (2) removal or reduction of phosphate from point and non-point sources. In case (1) phosphate 'builders' have either been banned from detergents or used sparingly. They have been replaced with substances like sodium nitrilotriacetate (NTA) and sodium citrate as well as zeolites. In case (2), PO_4^{3-} is removed from industrial and municipal waste before it is discharged into lakes or rivers. This is done by adding Ca^{2+} as lime, $Ca(OH)_2$, which precipitates the phosphate as $Ca_3(PO_4)_2$ and $Ca_5(PO_4)_3OH$. These are easily removed as sediment.

Nitrates in Water

Recall that nitrogen (N) can have a range of oxidation numbers (ON).

Table 9.3   Oxidation States of Nitrogen

ON	Compound	Property
-3	$NH_3(g)$, $NH_4^+(s)$	reducing
0	$N_2(g)$	
1	$N_2O(g)$ (nitrous oxide)	
2	$NO(g)$ (nitric oxide)	
3	$NO_2^-(s)$ (nitrite ion), HNO_2 (nitrous acid)	
4	$NO_2(g)$ (nitrogen dioxide)	
5	$NO_3^-(s)$, HNO_3	oxidizing

9 - Water Pollution

The lower oxidation states (NH_3, NH_4^+) are present in anaerobic (reducing) environments such as the bottom of lakes, whereas the highest oxidation state (NO_3^-) occurs in aerobic environments such as the surface of lakes. Intermediate states, such as NO_2^-, exist in anaerobic environments that are not as reducing, e.g. water logged soils. Most plants absorb nitrogen in the form of NO_3^-, so the NH_4^+ ion (applied as fertilizer) must first be oxidized to NO_3^- by nitrifying bacteria before it can be utilized by plants.

Natural water contains some dissolved NO_3^- but excess NO_3^- may enter waterways as run-off from agricultural lands. In those cases where phosphorus (P) is the limiting nutrient, the excess NO_3^- does not pose a serious problem. However, there are instances in which nitrogen and not phosphorus is the limiting nutrient and excess NO_3^- causes algal blooms that pollute the water when the algae die.

Excess NO_3^- in drinking water is also linked to a disease in infants called methemoglobinemia (blue-baby syndrome). Apparently, bacteria in unsterilized feeding bottles cause the reduction of nitrates to nitrites:

$$NO_3^- + 2\,H^+ + 2\,e^- \rightarrow NO_2^- + H_2O$$

NO_2^- then reacts with hemoglobin in the blood thus preventing the transfer of O_2 as oxyhemoglobin to cells. As a result, the baby turns blue and dies of respiratory failure.

NO_2^- (from NO_3^-) in the body of adults is suspected of causing stomach cancer, but this has not been definitely established. The NO_2^- is assumed to form carcinogenic compounds called nitrosamines by combination with amines in the body. Small amounts of these nitrosamines have also been found in drinking water and pose a definite health hazard.

9.4 Heavy Metals in Water

Certain metals near the bottom of the periodic table have densities that are large compared to other elements. Examples are mercury (Hg; d = 13.5 g/mL), lead (Pb; d =11.3 g/mL), cadmium (Cd; d = 8.7 g/mL) and arsenic (As; d = 5.8 g/mL). We will confine our discussion to Hg and Pb. These heavy metals are particularly toxic to lower organisms and even man. They are not usually toxic as free elements but as cations and organic compounds (a phenomenon called speciation). Their toxicity is

Heavy Metals in Water

due to their strong affinity for the sulfhydryl group (SH) which is present in certain enzymes. Once attached to the metal, the enzyme cannot function properly leading to serious health effects. The reaction is:

$$2\ R\text{-}SH + M^{2+} \rightarrow M\begin{matrix}\diagup S\text{-}R \\ \diagdown S\text{-}R\end{matrix} + 2\ H^+$$

The antidote to these metals is called British Anti-Lewisite (BAL) and contains -SH groups which undergo a similar, competitive reaction. Another factor which accounts for the acute toxicity of these ingested metals is that they bioaccumulate and biomagnify as they move up the food chain. Let us discuss briefly some of the chemistry of heavy metals, their uses and hence sources of the metals in the environment, and strategies for removing or at least reducing their presence in the environment. We will confine our discussion to two of these metals- mercury and lead.

Mercury

The symbol is Hg, the atomic number is 80 and the electronic configuration is: $[_{54}Xe]\ 6s^2\ 4f^{14}\ 5d^{10}$. It is a transition metal, occurring at the end of the second transition series.

Hg exists in two oxidation states: Hg_2^{2+} (+1) and Hg^{2+} (+2), of which +2 is the more stable, by loss of electrons:

$$Hg(s) \rightarrow Hg^{2+} + 2\ e^-$$

The free element is a shiny liquid at room temperature (the only liquid metal). You have seen it in mercury thermometers. It has an appreciable vapor pressure (for a metal) at room temperature (10^{-3} torr) and so the vapor which is highly toxic easily contaminates the air if the liquid is spilled (hence the necessity to clean up spills as soon as possible and to ventilate the area of use properly). The vapor can travel over long distances before being deposited on land or in the water.

The +2 oxidation state exists in such compounds as $HgCl_2$ (which is water soluble) and HgO, HgS (which are water insoluble). HgS is present in the ore, cinnabar, from which Hg is prepared by roasting in air:

9 - Water Pollution

$$HgS(s) + O_2(g) \rightarrow Hg(\ell) + SO_2(g)$$

Hg^{2+} is also found combined in organic compounds of the metal (organomercurials) which are used mainly as fungicides on things like seedlings and so may enter the environment in that way.

Uses/Sources

Liquid Hg is used in electric switches, in fluorescent light bulbs and Hg lamps (used for street lighting). Hg is also used in Hg batteries (as HgO), in cameras and hearing aids. Hg may enter the environment by incineration of solid waste containing Hg products. Further, large amounts of Hg may enter the environment when coal and fuel oil, containing trace amounts of the element, are burnt.

Many metals dissolve in liquid Hg to form a solution called an amalgam. A solid amalgam with tin (Sn) and silver (Ag) is used as dental filling. (some scientists believe that small amounts of Hg may leach out of these fillings and cause health problems). An amalgam of Na and Hg at one time was used as an electrode in the manufacture of Cl_2 and NaOH (chlor-alkali process) by electrolysis. Loss of Hg during the recycling process occurred when the plant's cooling water was returned to the river. Non- mercury electrodes are now being used.

Methylation of Hg

Much of the Hg that enters the environment does so either as elemental Hg or as Hg^{2+}. Either way, it eventually gets converted to 'methyl mercury' [CH_3HgX or $(CH_3)_2Hg$] by the action of certain anaerobic bacteria in sediments in the presence of a form of vitamin B_{12} called methylcobalamin, which contains cobalt. These methylated mercury compounds are extremely toxic because they are lipophilic (hydrophobic) and so dissolve in fat tissue and can then easily cross the blood-brain barrier.

Much of the Hg present in humans originate from fish in our food supply and are usually present as methyl mercury. [Just recently (1999) a chemistry professor at Dartmouth spilled $(CH_3)_2Hg$ on her hands. Even though she was gloved, she still died of Hg poisoning.] One of the most famous cases of Hg poisoning occurred in the village of Minamata in Japan in the 1950's. A chemical plant manufacturing PVC used a Hg^{2+} catalyst in the process and dumped the waste into the bay. The fish and shell fish in Minamata Bay bio-accumulated Hg and were eaten by the residents. Hg levels as high as 100 ppm were detected in the inhabitants (the threshold limit value for American diets is 0.5 ppm). About 1300 people came down with 'Minamata' disease-irritability, numbness in arms and legs, loss of hearing and sight, loss of muscle coordination and some 200 died.

Strategies of Control of Mercury

1. Elimination of Na/Hg electrodes in chloro-alkali plant
2. Use of British Anti- Lewisite as an antidote for Hg poisoning
3. Addition of soluble sulfide (S^{2-}) to precipitate Hg^{2+} as insoluble HgS.

Lead

The symbol for lead is Pb, the atomic number is 82 and the electronic configuration is $[_{54}Xe]\ 6s^2\ 4f^{14}\ 5d^{10}\ 6p^2$. Pb exhibits oxidation states of +2 and +4 by formally losing its $6s^2$ and $6p^2$ electrons respectively. For example, the +2 oxidation state is formed by the reaction:

$$Pb(s) \rightarrow Pb^{2+} + 2\ e^-$$

The predominant oxidation state is +2. The +4 state exists in such compounds as $PbO_2(s)$ which is a component of lead-acid batteries whereas the +2 state exists as $PbCl_2$ which is sparingly soluble in water and PbS, PbO which are water insoluble. PbS is the principal ore, called galena, from which Pb is extracted by first roasting in air:

$$2\ PbS(s) + 3\ O_2(g) \rightarrow 2\ PbO(s) + 2\ SO_2(g)$$

followed by reduction to the metal by heating with carbon.

$$2\ PbO(s) + C(s) \rightarrow 2\ Pb(s) + CO_2(g)$$

Uses/Sources

Pb is a dull grey metal which has been used extensively for plumbing since Babylonian times (about 2000 BC), partly because it is a relatively soft metal and so it is easy to work. Unfortunately, Pb metal itself may dissolve in water that is only slightly acidic and this is one way the metal

9 - Water Pollution

enters our drinking water.

$$Pb(s) + 2\ H^+(aq) \rightarrow Pb^{2+}(aq) + H_2(g)$$

Pb is also used in solder which is an alloy of Pb and Sn. It is also used in leaded glass for ornamental purposes. It is used as a shielding material in X-ray work, as a pigment in many paints, for example, lead chromate ($PbCrO_4$) in yellow paints (used on school buses and road signs) and basic lead carbonate, $2PbCO_3 \cdot Pb(OH)_2$ (a white pigment). Lead in these paints is a serious health hazard for children who might accidentally ingest the flakes of paint. However, most paints today are lead-free having been replaced by zinc oxide (ZnO) and titanium dioxide (TiO_2).

Pb is used extensively in lead-acid batteries which power automobiles and used car-batteries is an important source of lead in the environment, whether they are disposed of by just throwing them away in land fills or recycled by distillation.

Lead is sometimes added to low octane gasoline in the form of an organolead compound, tetraethyl lead (TEL), which serves as an anti-knock agent. Low octane gasoline causes knocking or pinging in the engine because of premature combustion of the gasoline/air mixture and reduces engine efficiency. TEL promotes smoother combustion of the gasoline by forming 'free radicals' on decomposition at the high temperature of the engine.

$$Pb(C_2H_5)_4(\ell) \rightarrow Pb(g) + 4\ C_2H_5 \cdot (g)$$

The Pb is removed as $PbCl_2$ by reaction with methylene chloride which is also added to the gasoline. Auto emissions from cars using leaded gasoline are a major source of lead in the environment (non-point source), not to mention emissions at the point of manufacture during the distillation of TEL (point source). Leaded gasoline has been banned in the United States of America and other developed countries but still finds its way into gasoline supplies in many developing countries.

The residence time of Pb in humans is about 6 years in whole body and 15-20 years in skeletal bones. However it can be removed fairly easily by a chelating agent such as EDTA (ethylenediaminetetra-acetate) and BAL. The threshold limit value (TLV) for Pb is 100 ppb.

Biochemical Oxygen Demand

9.5 Organic Matter in Water

Organic matter enters waterways from many sources-sewage (human/animal), factory effluents, agricultural run-off, dead plants, leaves. This organic matter is usually oxidized by dissolved oxygen (DO) in the presence of aerobic micro-organisms to CO_2 and H_2O.

$$(CH_2O)_n + n\, O_2 \rightarrow n\, CO_2 + n\, H_2O$$

so it is important that there is enough dissolved O_2 in the water to bring about this oxidation and still satisfy the needs of fish and other aquatic life. So let us talk about dissolved O_2 and a concept called the biological (biochemical) oxygen demand (BOD) of organic matter in H_2O.

Recall that gases are generally not very soluble in H_2O (unless they react chemically with H_2O, like NH_3). Their solubility is governed by a law called Henry's law which says that the solubility of a gas in a liquid is directly proportional to the partial pressure of the gas over the solution; that is, the higher the partial pressure, the higher the concentration of the solute gas. Or:

$$C_{O2} = K\, P_{O2}$$

At a total barometric pressure of 1 atm (in which $P_{O2} = 0.2$ atm) and 25 °C, the solubility of O_2 in water is estimated to be about 8.7 ppm. Recall also that the solubility of a gas decreases with increasing temperature so that at 0 °C the solubility of O_2 is 14.7 ppm and at 35 °C it is 7.0 ppm, so let us use an average value of 10 ppm at ambient temperatures. Fish need at least 5 ppm dissolved O_2 to survive so that any condition that causes the depletion of dissolved O_2 below 5 ppm is a serious threat to fish and other aquatic life. That is why organic matter like dead plants and animal/ human wastes in water pose a threat to fish life since they use up dissolved O_2 thereby reducing its availability to marine life.

The ability of dissolved organic matter in a sample of water to consume O_2 in the presence of aerobic micro-organisms is called its biochemical (biological) oxygen demand (BOD). It is measured by determining the O_2 used up by a suitable volume of the sample sealed in an air-tight container and incubated with aerobic bacteria at 25 °C for five days. (The volume is chosen so that all of the oxygen is not completely used up by the bacteria). A high BOD (for example, 15 ppm of O_2)

9 - Water Pollution

means that the organic matter can rapidly deplete the amount of dissolved O_2 available. The median BOD of unpolluted surface water is about 0.7 ppm

A faster, but less accurate, determination of dissolved O_2 demand is known as the chemical oxygen demand (COD) which is determined by using another substance known as potassium dichromate, $K_2Cr_2O_7$ (instead of O_2) as the oxidizing agent. COD is usually larger than BOD because $Cr_2O_7^{2-}$ will oxidize some organic matter that O_2 will not oxidize.

Water polluted by organic matter could have a BOD which exceeds 10 ppm (the maximum solubility of O_2 in water at ambient temperature) and unless the water is purposely re-aerated, no dissolved O_2 will be available and fish will die. That is why it is necessary to reduce the BOD of sewage waste water to a convenient level before releasing it into waterways.

Waste Water Treatment

The treatment of sewage waste water is divided into three phases- primary, secondary and tertiary. In the primary treatment stage, the water is first screened to remove large particles such as sticks and stones. It then enters a lagoon where the insoluble particles fall to the bottom as a sludge by a process called sedimentation and the liquid grease forms a surface layer which is skimmed off. Lime and aluminum sulfate may be added to help sedimentation by forming $Al(OH)_3$ which slowly settles and pulls the insoluble particles down with it. This primary treatment typically reduces the BOD by about 30%.

In the secondary treatment stage, much of the biological matter is oxidized in the presence of aerobic micro-organisms to CO_2 and H_2O. Air/O_2 is deliberately bubbled into the waste water. This secondary treatment reduces the BOD to about 100 ppm so that, on dilution with unpolluted H_2O, the discharged water now has a low enough BOD to support aquatic life.

An (optional) tertiary treatment stage involves the removal of certain deleterious chemicals, for example, phosphates which are not removed in the first two stages. The specific treatment depends on the chemical to be removed. For phosphate removal, lime, $Ca(OH)_2$, is added to precipitate the phosphate as calcium phosphate, $Ca_3(PO_4)_2$ which settles to the bottom of the holding tanks. At this stage also, a disinfectant such as chlorine gas or hypochlorite may be added to reduce the amount of pathogens before the water is discharged into the river.

9.6 Organic Compounds

Organic compounds are compounds containing carbon. Exceptions are the oxides of carbon - CO and CO_2 -, bicarbonates (HCO_3^-) and carbonates (CO_3^{2-}) which are usually considered inorganic compounds. Organics were first isolated from organisms (hence their name)and for a long time it was felt they could only be synthesized in organisms by a 'vital force' in the organism. Examples of these first organic compounds were carbohydrates, proteins and nucleic acids. The first man-made organic compound, urea, which was synthesized by Friedrich Wohler (1828) ended the 'vital force' theory.

To date, some 6 million organics have been synthesized including such substances as polymers (plastics), fibers, dyes, drugs and petroleum products. The large number of organic compounds is due to:

(1) the unique ability of C to bond covalently to each other in chains (a process called catenation) and to other atoms. Few atoms other than C have this ability; silicon can form a maximum of 8 Si-Si bonds, germanium about 5 Ge-Ge bonds and tin forms 2 Sn-Sn bonds.
(2) existence of isomers which are compounds with the same molecular formula but different structural formulas. As the number of C atoms increases, the possible isomers become greater and greater. For example, among a group of hydrocarbons called alkanes, C_8H_{18} has 18 isomers while $C_{10}H_{22}$ has 75 isomers.

Functional Groups

The study of the vast array of organic compounds is facilitated by organizing them into a relatively small number of classes or families which are characterized by the presence of ' functional ' groups in these classes. A functional group is defined as a small structural unit within a molecule that seems to undergo specific chemical reactions regardless of the nature of the rest of the molecule. It behaves more or less the same regardless of the compound in which it is present For example, ethanol, C_2H_5OH, contains the OH group (which is called the alcoholic group) and any compound containing the OH group behave somewhat similar to ethanol. For example, one chemical reaction of all alcohols is that they undergo oxidation. Moreover, a given compound may contain more than one functional group.

Some common functional groups and families are described briefly

9 - Water Pollution

below. We will only do a quick survey of these groups and mention one or two important compound(s) from each group.

Alkanes

Alkanes belong to the larger group of organic compounds called hydrocarbons, or compounds consisting only of C and H. They are obtained almost exclusively from fossil fuels - oil and natural gas. The alkanes contain single C-C and C-H bonds only and are said to be saturated Most alkanes (the so-called open-chain alkanes) have the general formula C_nH_{2n+2} where n = number of C atoms. For example, the first member is CH_4 (n=1) and is known as methane, C_2H_6 (n=2) is ethane and C_3H_8 (n=3) is propane.

Methane is the primary constituent of natural gas (as you recall, it is also produced by anaerobic bacteria in marshes, hence the name marsh gas). The other important alkanes in natural gas are propane and butane which are sold in pressurized containers in the form of liquids. Alkanes are relatively unreactive chemically. If an H atom is removed from an alkane, the resulting group is called an alkyl group and is generally represented by the symbol R. For example, CH_3 from methane is called the methyl group.

Alkenes

These are hydrocarbons which are characterized by the presence of carbon-carbon double bonds. Open chain compounds have the general formula C_nH_{2n}. The simplest alkene is C_2H_4 (n = 2) and is called ethylene. Ethylene is one of the most important organic chemicals. It is used in the manufacture of polyethylene (a polymer) as well as feed stock for the manufacture of other important compounds like ethanol and ethylene glycol (antifreeze).

The second member is C_3H_6 (n = 3) or propylene. Propylene is also used in the manufacture of polymer plastic (polypropylene). The third member is butene, C_4H_8 (n = 4) which exists as two isomers: 1- butene and 2- butene depending on the location of the double bond.

Alkynes

These are hydrocarbons containing the C≡C triple bond and have the general formula C_nH_{2n-2}. The simplest and most important member is

Functional Groups

acetylene, C_2H_2 (n = 2) which is extremely flammable and generates very high temperatures (about 2800 °C) on combustion. Because of that, it is used in oxyacetylene blow torches. The second member of the group is propyne, C_3H_4 (n = 3)

Alcohols

This family of compounds contain the -OH or hydroxyl group. The simplest member is methanol, CH_3OH and the second member is ethanol, C_2H_5OH. The third member is propanol, C_3H_7OH which exists as two isomers : 1- propanol (n-propanol) and 2-propanol (iso-propanol).

Methanol is an important solvent. It is also used as feed stock to make formaldehyde, another very important chemical. Ethanol is the alcohol in wine and other alcoholic beverages and is made by fermentation of sugars. It is used in gasohol (mixture of gasoline and alcohol). Oxygenated fuel cuts down on pollution. Isopropyl alcohol is present in rubbing alcohol.

Ethers

Ethers contain the -C-O-C- bond. The simplest member is dimethyl ether (CH_3OCH_3) but by far the most important ether is the second member, diethyl ether ($C_2H_5OC_2H_5$) which is simply called 'ether'. Ether was used in surgery as an anesthetic but has now been replaced by more sophisticated compounds. It is a 'low boiling' liquid (BP= 34.5 °C) and hence evaporates quickly.

Aldehydes and Ketones

This class of organic compounds contains the C=O or carbonyl group. If one (or both) of the two other bonds to C is to the H atom, the compound is an aldehyde, RCHO. The aldehydic group (-CHO) is present in some sugars, for example, glucose. If both of the two other bonds to the carbonyl group are to C atoms, then the compound is a ketone.

The simplest aldehyde is formaldehyde (HCHO) which is a gas at room temperature (BP: -21 °C). It is a very important starting material for manufacture of other useful materials, most notably the plastic, Bakelite. Bakelite is a hard plastic used for making utensils and formica, among other things. The second member is acetaldehyde, (CH_3CHO), a low boiling liquid (BP: 21 °C).

The simplest ketone is acetone which is arguably one of the most important ketones. It is very useful as a solvent and is present in a number of useful products like paints, nail polish and the like.

Carboxylic Acids

This family contains the -COOH group with an alkyl group R (or H) occupying the fourth position around C. The first member is HCOOH (formic acid) which is present in ants. The second member is acetic acid, CH_3COOH and is the acid found in vinegar.

Esters

Esters are derivatives of carboxylic acids in which the H of the -OH in the - COOH group is replaced by an alkyl group R. For example, ethyl acetate, $CH_3COOC_2H_5$, is an ester of acetic acid (R=ethyl). Many fragrances from flowers and fruits are esters, for example, octyl acetate in oranges.

Amines

Amines are organic derivatives of ammonia, NH_3, in which one, two or all three of the H's are replaced in turn by alkyl groups to form primary (RNH_2), secondary (R_2NH) and tertiary (R_3N) amines respectively. Like NH_3, they are Bronsted bases and their aqueous solutions are basic (pH > 7) due to the reaction:

$$RNH_2 + H_2O \rightarrow RNH_3^+ + OH^-$$

Examples are CH_3NH_2 (methylamine), $(CH_3)_2NH$ (dimethylamine) and $(CH_3)_3N$ (trimethylamine).

Amides

Amides are derivatives of carboxylic acids in which the -OH of the -COOH group is replaced by the $-NH_2$ group (the H's in $-NH_2$ could also be substituted). The result is a compound containing the -CONH- group which is also present in proteins when it is then called the **peptide** linkage. The simplest members are $HCONH_2$ (formamide) and CH_3CONH_2 (acetamide).

Halides

Halides are derivatives of hydrocarbons in which one or more H's have been replaced by a halogen. For example, CH_3Cl is derived from methane and is called methyl chloride.

Aromatics

Aromatics are organic compounds containing the benzene ring. The parent compound benzene, C_6H_6, itself contains all six C atoms joined together in the form of a six-membered ring. An important derivative of benzene is toluene, $C_6H_5CH_3$ in which an H in benzene is replaced by a -CH_3 group. It has replaced benzene itself as a useful solvent because of benzene's carcinogenicity.

An environmentally important class of aromatics is polycyclic aromatic hydrocarbons (PAHs) in which a number of benzene rings are joined together. The simplest members are naphthalene and anthracene. The members with four or more rings are very mutagenic and some are carcinogenic as well.

9.7 Organics in Water

There are many organic pollutants in water. However we will emphasize here just a few of them. Included in our discussion will be organics such as certain pesticides, dioxin and PCBs.

Pesticides

Pesticides are defined as substances that kill or control the spread of unwanted organisms. Most of them work by blocking some vital metabolic process in the organism. They fall into several categories depending on the type of organism affected.

(a) Insecticides are compounds that kill insects
(b) Herbicides kill plants
(c) Fungicides kill fungi

Pesticides were once considered a blessing until their environmental impact became apparent. Perhaps the first person to bring the problem of pesticide contamination of the environment to public attention was the

9 - Water Pollution

naturalist, Rachel Carson, with the publication of her book in 1962 called " Silent Spring ". In the book, she discussed the negative effects of pesticides like DDT.

Pesticides enter waterways either directly as in the spraying of mosquitoes (point source) or as run-off from farmland. (non-point source)

DDT

DDT is the acronym for dichlorodiphenyltrichloroethane. It was first synthesized in 1874 and was introduced during World War II as a very effective, relatively inexpensive pesticide. Even though it has been banned for use in the United States since 1972, it is still being used in some Third World countries, in spite of its problems in the environment because it is relatively inexpensive and effective.

DDT contamination wipes out populations of birds such as the bald eagle. When ingested by birds, DDT first gets converted to the compound, dichlorodiphenyldichloroethane (DDE) during metabolism, by loss of HCl. DDE is the active metabolite and interferes with an enzyme that regulates Ca^{2+} uptake in cells. The net result is that the affected birds produce eggs with shells that contain less calcium and therefore are very fragile. Many of these eggs break during the hatching of their offspring and leads to a reduction in their population.

DDT is fairly toxic. Its toxicity is partly due to the fact that it bioaccumulates in organisms because of its hydrophobicity. Hydrophobic (non-polar) compounds like DDT are very fat soluble (lipophilic) and tend to accumulate in fatty tissue at concentrations in excess of that present in surrounding water. Biomagnification also occurs as we move up the food chain.

A control strategy to reduce the negative effects of DDT while preserving its usefulness as a pesticide would be to synthesize an analog of DDT that does not bioaccumulate. One such analog is **methoxychlor** in which the basic DDT structure has been slightly modified (a Cl atom is replaced by an -OCH_3 group).

DDT has been replaced by other pesticides such as organophosphates which are organic compounds containing phosphorus; for example, malathion. Organophosphates are less persistent in the environment due to hydrolysis and are more selectively toxic to insects.

Dioxin

During the Vietnam war, the herbicide known as agent orange was used as a defoliant by the United States armed forces in order to see the enemy more clearly in the jungle. Agent orange is a mixture of two pesticides: 2,4-D (dichlorophenoxyacetic acid) and 2,4,5-T (trichlorophenoxyacetic acid). One of the by-products of 2,4,5-T synthesis is an extremely toxic substance known as dioxin.. There are a number of dioxins, but the most common and the most dangerous is tetrachlorodibenzo-dioxin (TCDD).

Other sources of dioxin are:

1. Burning of tetra- and penta-chlorophenol (PCP)-treated wood. PCP is used as a herbicide.
2. Bleaching of pulp and paper by Cl_2 or ClO_2. Dioxin may result from chlorination of organics in the pulp.
3. Incineration of organochlorine compounds.

The release of dioxin into our waterways creates a very serious environmental problem. Because of its hydrophobicity, it bioaccumulates in fish and animals and can finally enter food for humans. It is known to affect reproduction in rats and is also carcinogenic to some animals.

PCBs

PCB is the acronym for polychlorinated biphenyls. They are manufactured by the chlorination of the parent compound - biphenyl. They have unique properties. For example, they are water insoluble and lipophilic. They are chemically inert and non-biodegradable liquids with low vapor pressure. These properties make them very useful compounds in society. For example, PCBs are coolants in power transformers and capacitors. They are also used as plasticizers in PVC plastic tubing to make them more flexible, in 'carbonless' copy paper, in waterproofing, and as de-inking solvents in newspaper recycling.

These same properties make them environmental hazards when they enter our waterways by improper disposal because they are very persistent in the environment. The Hudson River in New York became contaminated with PCBs as a result of discharge of waste from capacitor producing plants. The PCBs sink to the bottom of the river and become entrapped in

9 - Water Pollution

the sediment there. Some of the environmental consequences of PCBs are:

1. PCBs are persistent in the environment because of their non-biodegradability. They bioaccumulate and biomagnify.
2. They affect human health even though they are not acutely toxic. High doses in rats have been found to be carcinogenic.
3. Exposure to PCBs causes a severe skin condition called chloracne-a disfiguring analogue of common acne.
4. They may affect reproduction in animals and also seem to affect growth in children.

Their toxicity in animals may be due to metabolites of PCBs rather than the compounds themselves.

Control Strategies

Because PCBs represent a widespread and persistent problem in the environment, their use in transformers and capacitors was banned in the United States in 1973. Some substitutes have been developed to replace PCBs in electrical equipment. The disposal of the remaining PCBs is a difficult problem, again because of their chemical and thermal stabilities. The best method seems to be incineration at very high temperatures (about 1200 °C) in the presence of oxygen; they are converted to CO_2, H_2O and HCl during incineration.

SELECTED REFERENCES

Section 9.1 - Distribution of Water

1. J. M. Beard, "Chemistry, Energy and the Environment", pp 173 - 190, Wuerz, 1995
2. C. Baird, " Environmental Chemistry", pp 287 - 290, 302 - 309, Freeman, 1995

Section 9.2 - 9.3 Phosphates and Nitrates in Water

3. J. M. Beard, ibid., pp 193 - 199
4. C. Baird, ibid., pp 290 - 294, 310 - 312

Section 9.4 Heavy Metals in Water

5. C. Baird, ibid, pp 347 - 376

Section 9.5 Organic Matter in Water

6. J. M. Beard, ibid., pp 210 - 217
7. C. Baird, ibid., pp 294 - 300, 315 - 317

Section 9.6 Organic Compounds

8. C. Baird, ibid., pp 193 - 212

Section 9.7 Organics in Water

9. C. Baird, ibid., pp 215 - 231, 244 - 270

ADDITIONAL REFERENCES

1. S. E. Manahan, " Fundamentals of Environmental Chemistry ", Lewis Publishers, 2001
2. S. E. Manahan, " Environmental Chemistry", Lewis Publishers, 2000
3. N. J. Bunce, " Introduction to Environmental Chemistry", Wuerz Publishing, 1993
4. N. J. Bunce, " Environmental Chemistry ", Wuerz Publishing, 1994
5. G. T. Miller, " Environmental Science ", Wadsworth Publishing, 1999
6. D. B. Botkin, E. A. Keller, " Environmental Science ", Wiley, 1998
7. P. H. Raven, L.R. Berg, G. B. Johnson, " Environment ", Saunders, 1998

QUESTIONS

Section 9.1 - Distribution of Water

9.1   Describe three regions of the earth's surface that contain water and give their approximate abundances.

9 - Water Pollution

9.2 Describe briefly the main steps in the purification of drinking water.

Section 9.2 - Water Pollutants

9.3 Define water pollution. Describe three ways in which water may become polluted.

Section 9.3 - Phosphates and Nitrates in Water

9.4 Describe three ways in which water systems may be contaminated with phosphate.

9.5 What is eutrophication and what is its origin?

Section 9.4 - Heavy Metals in Water

9.6 Name two metals which pollute our water. For each metal, list two ways by which the metal enters the environment.

9.7 Write a short essay on mercury and its effect on the aquatic environment.

9.8 Write a short essay on lead and its effect on the aquatic environment.

Section 9.5 - Organic Matter in Water

9.9 Oxygen dissolves in water to a small extent. How does its solubilty depend on the temperature of the water and the pressure of the oxygen?

9.10 Name three sources of organic matter in water. What happens to the organic matter in the presence of dissolved oxygen?

9.11 Define the term biological (or biochemical) oxygen demand (BOD) as applied to organic waste in water. How does it differ from the chemical oxygen demand (COD) of the waste?

Questions

9.12 Describe the primary, secondary and tertiary stages of waste water treatment. Write equations where possible.

Section 9.6 - Organic Compounds

9.13 Define an organic compound. Give two examples.

9.14 What is meant by a functional group in organic chemistry? Name two functional groups and write the structural formula of a compound containing each group.

9.15 Name the functional group in each of the following structural formulas. Name also the compound. (a) $CH_3CH=CH_2$ (b) CH_3Cl (c) CH_3CH_2COOH (d) CH_3COCH_3

Section 9.7 - Organics in Water

9.16 What does the acronym DDT stand for? Describe why DDT in a pond poses a serious health hazard to birds that eat fish from the pond.

9.17 What is Agent Orange and why is it a threat to humans?

9.18 What does the acronym PCB stand for? List three properties of PCB's that made them very useful compounds before they were banned.

9.19 Name two organics in water that pose a risk to humans. Describe briefly a strategy for reducing the risk in each case.

Chapter 10 - Human Health and the Environment

10.1 Chemical Toxicology

As you are aware, there are many substances in the environment that are hazardous to human health. Hazardous chemicals are defined as chemicals that pose a serious threat to human health or the environment if they are handled or disposed of improperly. Hazardous chemicals are of different kinds depending on their chemistry but we shall be concerned mainly with those which are classified as toxins.

Many chemicals are hazardous to your health. In fact, most, if not all substances, are hazardous at high enough concentrations. For example, potassium chloride, KCl is a relatively harmless substance under normal conditions. But it becomes Dr. Kevorkian's assisted suicide chemical at high concentrations. However, there are some substances which are harmful to your health at relatively low concentrations (ppm) and these are called toxins. The study of toxins and their environmental effects is called toxicology. Specifically, we will be concerned here with chemical toxicology in which the toxins are chemicals rather than micro-organisms.

Measurement of toxicity

Generally, the effects of a substance on a particular organism depend on the concentration or **dose** of the substance. This dose dependence can be represented by a dose-response curve in which a stated response in the test animal is studied and plotted as a function of concentration of the substance, usually mg of toxin per kg body weight of animal (written as mg/kg). Because individual animals, even of the same species, may vary in their responses, a large number of animals is usually tested in order to obtain a statistically significant response.

A quantity called the effective dose (ED) measures the % of the animal population displaying a particular response (e.g. a rash, loss of hearing, slurred speech, nausea, death) at increasing doses of the substance. The ED_{50} is the dose at which 50% of the population shows the given response. When the response is death of the animals, the ED_{50} is called the LD_{50} (lethal dose) as illustrated in Figure 10.1. LD_{50} is usually expressed as mg/kg.

The smaller the LD_{50}, the more toxic the substance to the particular animal. Note that LD_{50} for a given toxin may vary from species to species. It may also depend on the method of administering the toxin

Chemical Toxicity

which may be done subcutaneously, orally or intravenous (IV). Thus, when an LD_{50} is quoted for a toxin, it is important to specify the test animal as well as the method of administration. Thus, the LD_{50} of dioxin is

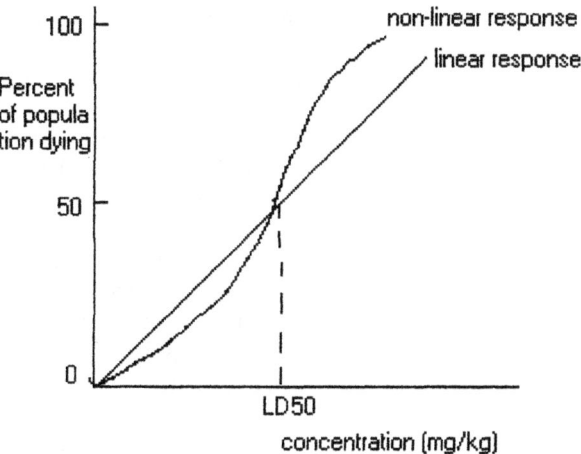

Figure 10.1 The definition of LD50 of a toxin

1.6×10^{-3} mg/kg for guinea pigs but 3.0 mg/kg for hamsters (presumably administered subcutaneously). This shows also that one has to be careful in extrapolating toxicity data on animals to humans. However, because of the difficulty of getting toxicity data on humans, the animal data are often useful in predicting **probable** behavior in humans.

10.2 Classification of Chemical Toxins

Toxic substances are classified according to their mode of action physiologically into (1) corrosive toxins (2) metabolic toxins (3) teratogens (4) mutagens (5) carcinogens. We shall briefly discuss each category using examples of environmental contaminants covered previously in the text.

10 - Human Health and the Environment

1. Corrosive Toxins

Corrosive toxins or irritants act by destroying,, on contact, the proteins of biological tissues. Some may attack skin and muscle tissue, others may attack the mucous membrane of the mouth and throat, still others may attack lung tissue. The most important environmental irritants are sulfur dioxide (SO_2), nitrogen dioxide (NO_2)and ozone (O_3). The reactions of irritants are irreversible and the only solution is removal of the offending pollutant.

SO_2 is a respiratory tract irritant. It damages lung tissue, causing constriction of the airways and therefore making breathing difficult. SO_2 is lethal at 500 ppm which is much greater than the 10 ppm level normally encountered in polluted air. However even at 10 ppm, SO_2 will cause distress to asthmatics and people suffering from other respiratory diseases such as emphysema. Plants are also sensitive to SO_2 pollution.

NO_2, a component of photochemical smog, can also cause lung inflammation although it is not as serious as sulfur dioxide.

O_3, also a component of photochemical smog,, is a powerful oxidant which destroys tissues by oxidizing the lipid coatings on cell walls in the lung, and deactivating enzymes. The victim experiences coughing,, shortness of breath, nose and throat irritation. Plants are also adversely affected.

2. Metabolic Toxins

Metabolic toxins act by interfering with some essential metabolic process. They include environmental contaminants such as carbon monoxide (CO), heavy metals and pesticides.

CO, a major component of automobile exhaust, acts as a powerful toxin because of its great affinity for the hemoglobin of the blood. Hemoglobin (Hb) is a protein containing Fe (II) and is responsible for carrying O_2 to tissues by first combining with O_2 in the lungs to form oxyhemoglobin.

$$Hb + O_2 \rightleftharpoons HbO_2$$

The reaction is reversible and the HbO_2 releases O_2 to the tissues as needed. However in the presence of CO, hemoglobin forms carboxyhemoglobin.

Metabolic Toxins

$$Hb + CO \rightarrow HbCO$$

CO has an even greater affinity than O_2 for hemoglobin and so cells are effectively starved of O_2. If the victim is administered pure O_2 quickly, the CO reaction could be practically reversed and so bring some relief to the victim.

Heavy metals like Hg and Pb act by deactivating enzymes. Enzymes contain the sulfhydryl (-SH) group with which heavy metals react.

Hg poisoning is characterized by jerking, irritability, mental instability. An antidote to Hg poisoning is a compound developed by the British during World War I to counter the effects of an As-containing chemical warfare agent called Lewisite (after its discoverer W. Lee Lewis). The antidote is therefore called British Anti Lewisite or BAL. It also contains the -SH group and therefore reacts the same way with Hg^{2+} as the enzyme does and therefore competes effectively with the enzyme for the Hg^{2+}.

Pb^{2+} can also be removed from the body by BAL and another compound called EDTA (ethylenediamine tetra-acetate) which is added as the Ca^{2+} salt.

$$(CaEDTA)^{2-} + Pb^{2+} \rightarrow (PbEDTA)^{2-} + Ca^{2+}$$

Pesticides such as parathion and malathion are neurotoxins because they attack the central nervous system. Parathion is an organophosphate insecticide and is a 'cousin' of the nerve gas Sarin, which you may have heard about in connection with the Gulf War. Neurotoxins act by disrupting the transmission of nerve signals.

3. Teratogens

Teratogens are substances which induce birth defects when a developing embryo is exposed to them, especially during the first two months of pregnancy. The most infamous teratogen is probably thalidomide which was used in Germany and Britain during the 1960's as a tranquilizer and sleeping pill. Women who took the drug during pregnancy at times gave birth to babies without arms and legs. Strangely enough, thalidomide is now being tested as an anti-cancer agent. From the point of view of the environment, compounds such as PCBs, Pb and Hg compounds are established teratogens.

10 - Human Health and the Environment

4. Mutagens

Mutagens are substances that alter the sequence of bases in DNA, the genetic material of cells. The process is called mutation. When these mutations occur in the DNA of germ cells (egg and sperm cells), they could lead to birth defects or to the transmittance of hereditary diseases like cystic fibrosis. An environmentally important class of mutagens is a group of organic compounds called polycyclic aromatic hydrocarbons (PAH's), of which benzopyrene is an example. The food preservative, sodium nitrite ($NaNO_2$) is also a suspected mutagen.

5. Carcinogens

Carcinogens are componds that cause cells to grow abnormally and uncontrollably, thereby producing a malignant tumor. The exact mechanism is unknown but it is assumed that the initial event is damage to the DNA of cells. Not surprisingly, some mutagens are also carcinogens, for example, benzopyrene mentioned earlier. The occurrence of certain cancers seems to correlate with certain factors in the environment. For example, mineralized H_2O containing high NO_3^- and SO_4^{2-} levels seems to correlate with an increased occurrence of stomach cancer.

SELECTED REFERENCES

Section 10.1 - 10.2 Chemical Toxicology

1. J. M. Beard, " Chemistry, Energy and the Environment ", pp 65 - 84, Wuerz, 1995
2. D. B. Botkin, E. A. Teller, " Environmental Science ", pp 297 - 304, Wiley, 1998

ADDITIONAL REFERENCES

1. P. Buell, J. Girard, " Chemistry - An Environmental Perspective ", Prentice Hall, 1994
2. S. E. Manahan, " Fundamentals of Environmental Chemistry ", Lewis Publishers, 2001

QUESTIONS

Section 10.1 - Chemical Toxicology

10.1 What is meant by a hazardous chemical? What are chemical toxins?

10.2 Define LD_{50}. What do we mean when we say the LD_{50} of dioxin in rats is 0.022 mg/kg? What is missing from the statement?

Section 10.2 - Classification of Chemical Toxins

10.3 What is a teratogen? Give two examples.

10.4 What is a mutagen? Give two examples.

10.5 What is a carcinogen? Give two examples.

10.6 What is the difference between a mutagen and a carcinogen? Are all mutagens carcinogenic and are all carcinogens mutagenic?

Appendix

Appendix

Selected Videos on the Environment

1. The Earth at Risk Video Series by Schlessinger Productions

   . Acid Rain
   . Clean Air
   . Clean Water
   . Degradation of the Land
   . Global Warming
   . Nuclear Energy/Nuclear Waste
   . The Ozone Layer

2. Introduction to Environmental Technology by Intelecom

   . Paths of Pollution (Part 6)

3. Chemical Cycles in the Biosphere - Hawkhill Videos

4. Sewage Treatment - Classroom Video

5. Race to Save the Planet Series - Annenberg, 1990

   . Do We Really Want To Live This Way (Part 3)
   . More For Less (Part 6)
   . Waste Not, Want Not (Part 8)

Index

Absolute temperature, 12
Acetaldehyde, 137
Acetic acid, 112, 138
Acetone, 138
Acetylene, 137
Acid-base indicators, 115
Acid-base theory, 108
Acid deposition, 87
Acid rain, 87
Acids, 108
  Arrhenius, 108
  Bronsted-Lowry, 108
  carboxylic, 138
  pH of, 113
  strength of, 111
  strong, 120
  weak, 120
Activation energy, 53
Aerobic bacteria, 71, 133
Aerosols, 84
Agent Orange, 141
Agriculture, 2
  fertilizers, 126
  pesticides, 139
Air, 66
  polluted, 77
  unpolluted, 67
Air pollution, 77
  photochemical smog,, 77
  stratospheric ozone depletion, 82
  acid rain, 87
  global warming,, 91
Alcohols, 137
Aldehydes, 137
Alkali metals, 44
Alkanes, 136
Alkenes, 136
Alkynes, 136

Aluminum, 10
Amalgam, 130
Amides, 138
Amines, 138
Ammonia, 57, 95
Ammonium sulfate, 126
Anaerobic bacteria, 130, 136
Anion, 111
Anti-knock agent, 132
Aqueous solutions, 104
Aquifer, 123
Argon, 73
Aromatics, 139
Arrhenius, 108
Arsenic, 128
Asthenosphere, 6
Atmosphere, 67
Atom, 18
Atomic mass, 20
Atomic mass unit, 20
Atomic number, 19
Atomic structure, 22
Atomic theory, Dalton's, 18
Avogadro's law, 64
Avogadro number, 34

Bacteria, 130, 133
Bases, 108
Bioaccumulation, 129
Biomagnification, 129
Birth defects, 149
Blue-green algae, 71
Bohr, Niels, 22
Boiling point, normal, 102
Boiling point elevation, 107
Bond energy, 36
Bonding, 35
  covalent, 36
  ionic, 35

153

Index

Boyle's law, 64
BAL, 129, 149
Bronsted-Lowry, 109
  acids, 109
  bases, 109
Buffers, 116
Butane, 136
Butene, 136

Cadmium, 128
Calcium carbonate, 72
Calcium hydroxide, 134
Calcium oxide, 70, 89
Calcium phosphate, 126, 134
Calcium sulfite, 89
Calcium sulfate, 89
Calorie, 12
Cancer, 128, 150
Carbohydrates, 71, 135
Carbon, 21, 48
Carbon cycle, 72
Carbon dioxide, 72
  greenhouse effect, 93
Carbon monoxide, 78, 148
Carbonyl group, 137
Carboxyhemoglobin, 148
Carboxylic acids, 138
Carcinogens, 150
Catalyst, 53, 58
Catalytic converter, 81
Cation, 111
Celsius temperature, 12
Chain reaction, 85
Charles' law, 64
Chemical bond, 33
Chemical change, 12, 46
Chemical equations, 46
Chemical equilibrium, 55
Chemical formulas, 48

Chemical property, 12
Chemical reactions, 47
Chemical symbol, 18
Chile saltpeter, 69
Chlorine, 124, 134
Chlorofluorocarbon, 84
Coal, 89
Colligative properties, 107
Combination reaction, 48
Combustion, 46
Compounds, 11
Concentration, 52, 105
  and reaction rates, 52
Condensation, 101
Conservation of energy, 51
Conservation of mass, 18
Constant composition, law of, 18
Conversion factors, 14
Core, earth's, 6
Core electrons, 26
Corrosive toxins, 148
Covalent bond, 35
Crust, earth's, 6

Dalton, 18
  atomic theory, 18
  law of partial pressures, 66
DDT, 140
Definite proportions, law of, 18
Denitrification, 95
Density, 10
Desalination, 122
Detergents, 126
Deuterium, 20
Diatomic molecules, 37
Dichlorophenoxyacetic acid, 141
Diethyl ether, 102, 137
Dimensions, 13
Dimethyl ether, 33, 137

Index

Dioxin, 141, 147
Dipole, 38, 100
Diseases, 124, 148
Dissociation energy, 68
Dissolved oxygen, 133
Distillation, 132
DNA, 86, 150
Dose-response curve, 146
Double bond, 39
Drinking water, 123
Dynamic equilibrium, 55

Earth, 1, 3, 6
EDTA, 132, 149
Electrolyte, 35
Electromagnetic radiation, 91
Electron, 19
Electronic configuration, 25
Electron dot structure, 30
Electronegativity, 37
Electrostatic attraction, 35
Elements, 11, 18
Endothermic reaction, 51
Energy, 49
Environment, 1
Enzymes, 54, 129
Equations, chemical, 46
Equilibrium, chemical, 55
Esters, 138
Ethanol, 135, 137
Ethers, 137
Ethyl acetate, 138
Ethylene, 136
Ethylene glycol, 108, 136
Eutrophication, 127
Exothermic reaction, 51

Fahrenheit scale, 12
Fermentation, 137

Fertilizers, 126
Fish, 133
Fluorine, 84
Food chain, 129
Formaldehyde, 80, 137
Formic acid, 138
Formulas, molecular, 33
Fossil fuels, 72, 136
Freezing point, 108
Functional groups, 135
Fungicide, 139
Fusion, heat of, 103

Gases, 62
  laws, 64
  real, 66
Gasoline, 77, 132
Giant planets, 4
Global warming, 91
Glucose, 107, 137
Graphite, 22
Greenhouse effect, 91
Greenhouse gases, 91
Ground state, 25
Ground water, 123
Groups, functional, 135
  periodic table, 21

Haber process, 70
Halons, 85
Hard water, 124
Hazardous chemical, 146
Heat, 12
Heat capacity, 49
Heat of fusion, 103
Heat of vaporization, 102
Helium, 18, 25
Hemoglobin, 148
  carboxy-, 148

Index

Hemoglobin, oxy-, 148
Herbicide, 139
Heterogeneous catalyst, 54
  mixture, 11
Homogeneous catalyst, 54
  mixture, 11
Hydrocarbons, 136
  alkanes, 136
  alkenes, 136
  alkynes, 136
  aromatic, 139
Hydrochloric acid, 112
Hydrogen, 8, 19, 22
Hydrogen bond, 100
Hydrogen chloride, 37
Hydrogen ion, 108
Hydrogen peroxide, 71
Hydrogen sulfide, 67
Hydrologic cycle, 74
Hydronium ion, 109
Hydroxide ion, 110
Hydroxyl radical, 80

Ice, 103
Incineration, 130, 142
Inert gases, 36
Infrared radiation, 91
Insecticide, 139, 149
Intermolecular forces, 100
International system of units, 13
Ionic bond, 35
Ionic compounds, 35
Ions, 44
Iron, 6
Irrigation, 122
Isomers, 33, 135
Isotopes, 19

Joule, 12, 50

Kelvin scale, 12
Ketones, 137
Kilogram, 14
Kinetic energy, 49

Lakes, 122
Lavoisier, 18
Law of conservation of mass, 18
Law of definite proportions, 18
Law of multiple proportions, 18
Law of thermodynamics, first, 50
LD_{50}, 146
Lead, 131
Leaded glass, 132
Lethal dose, 146
Lewis, G. N., 36
Lewisite, 149
Lewis structure, 36
  dot structure, 29
Lightning, 70
Lime, 90, 127
Limestone, 73, 90
Liquids, 11, 100
Lithium, 68
Lithosphere, 1, 6
Litmus, 115
London forces, 101
Lower atmosphere, 68
Lowry, 109

Magnesium, 7
Magnesium chloride, 111
Magnesium oxide, 35
Magnetic field, 91
Malathion, 140, 149
Malignant tumor, 150
Mantle, 1, 6
Mars, 5
Mass, 10

Mass number, 19
Matter, 10
Melting point, 11
Mendeleev, 21
Mercury, 129
Mercury electrodes, 130
Mercury (II) oxide, 51
Mesosphere, 68, 82
Metabolic toxins, 148
Metalloids, 22
Metals, 21
Methane, 39, 93
Methanol, 137
Methemoglobinemia, 128
Methylamine, 138
Methyl chloride, 139
Methyl mercury, 130
Mixtures, 11
Molarity, 106
Molar mass, 34
Mole, 34
Molecular mass, 33
Molecular structure, 12
Molecules, 33
Montreal protocol, 87
Multiple proportions, law of, 18
Mutations, 150
Mutagens, 150

Natural gas, 46, 94
Neon, 26
Neurotoxins, 149
Neutralization, 109
Neutrons, 19
Newton, 10
Nickel, 6
Nitrates, 127
Nitric acid, 81
Nitrification, 95

Nitrites, 128
Nitrogen, 68
Nitrogen fixation, 69
Nitrogen oxides, 77
Nitrosamines, 128
Noble gases, 36
Nomenclature, 45
Non-bonding electrons, 37
Non-metals, 22
Non-point sources, 125
Non-polar molecules, 37
Nucleus, of atoms, 19
Nutrient cycles, 68

Oceans, 121
Octane, 132
Octet rule, 39
Orbital, 26
Organic compounds, 135
Organic matter, 133
Organophosphates, 140
Osmosis, reverse, 124
Oxidation, 47
Oxides, 70
Oxygen, 70
Oxygen cycle, 71
Ozone, 83
Ozone hole, 85

PAHs, 139, 150
Palladium, 81
PAN, 78
Parathion, 149
Parts per million, 106
Pascal, 63
Pathogens, 125
Pauling, 37
PCBs, 141
Peptide linkage, 138

Index

Periodic table, 21
Periods, 21
Peroxyacetyl nitrate, 78
Pesticides, 139
pH, 113
Phosphate, 125
Phosphoric acid, 125
Photochemical smog, 77
Photosynthesis, 71
Physical change, 12
Physical property, 12
Platinum, 81
Plumbing, 131
Point source, 125
Poisoning, 130, 149
Polar bond, 37
Polychlorinated biphenyls, 141
Polycyclic aromatic hydrocarbons, 139, 150
Polyphosphate, 126
Potassium, 7
Potassium chloride, 146
Potassium dichromate, 134
Potential energy, 49
Precipitation, 74
Pressure, 63
Primary pollutant, 78
Primary amine, 138
Primary sewage treatment, 134
Principal quantum number, 24
Products, 46
Propane, 136
Propanol, 137
Propene, 136
Properties, 12
Propyne, 137
Protein, 135, 138
Protons, 19
Pure substance, 11

Quantum numbers, 24
Quartz, 90

Radiation, 91
Radical, 80
Rain, 87
Rainwater, 116
Rate of reactions, 51
Reactants, 46
Recycling, 141
Redox reaction, 47
Reduction, 47
Representative elements, 21
Respiration, 71
Reverse osmosis, 124
Rhodium, 81
Ring, benzene, 139
Rutherford, 22

Salinity, 122
Saliva, 116
Salts, 111
Saturated hydrocarbons, 136
Saturated solution, 105
Scientific notation, 14
Scrubbers, 90
Sea water, 121
Secondary amine, 138
Secondary pollutants, 81
Secondary sewage treatment, 134
Sewage, 125
Sewage treatment, 134
Significant figures, 14
Silicon, 7
Single bond, 36
Sludge, 134
Smog, 77
Soaps, 126

Index

Sodium, 35, 52
Sodium carbonate, 45, 124
Sodium chloride, 35, 45
Sodium nitrate, 69
Sodium nitrite, 150
Softening of water, 124
Solar radiation, 83
Solar system, 3
Solids, 10
Solubility, 105
Solute, 104
Solutions, 104
Solvent, 104
Sommerfeld, 24
Soot, 77
Stability, 36
Star, 4
States of matter, 10
Stratosphere, 82
Strengths of acids, 111
Strengths of bases, 111
Structural formula, 33
Subatomic particles, 19
Sulfhydryl group, 129
Sulfur, 89
Sulfur dioxide, 89
Sulfuric acid, 89
Sun, 4
Sunscreens, 86
Surface water, 122
Symbols, 18

2,4,5-T, 141
TCDD, 141
Temperature, 12
Teratogens, 149
Terrestrial planets, 4
Tertiary amine, 138
Tertiary sewage treatment, 134

Tetraethyl lead, 132
Thalidomide, 149
Theory, Dalton's atomic, 18
  kinetic, 62
Thermodynamics, first law of, 50
Thermosphere, 68
Thomson, J. J., 19
Toluene, 139
Toxicology, 146
Toxins, 146
Transition elements, 22
Triple bond, 40
Troposphere, 68

Ultraviolet radiation, 80, 86
Units, 13
Unsaturated solution, 105
Unshared electron pairs, 37
Upper atmosphere, 68
Urea, 135
UV radiation, 86

Valence electrons, 27
Van der Waals forces, 101
Vaporization, 101
Venus, 4
Vinegar, 106, 138
Volcanoes, 67
Volume, 14

Waste, organic, 125
Waste water treatment, 134
Water, properties, 100
Water pollution, 125
Water softening, 124
Water vapor, 101
Watt, James, 2
Wave mechanics, 24
Weak acid, 112

Index

Weak bases, 112
Weight, 10
Work, 12

X-radiation, 92

Zeolites, 126
Zinc oxide, 132

www.ingramcontent.com/pod-product-compliance
Lightning Source LLC
Chambersburg PA
CBHW072136160426
43197CB00012B/2130